D0998550

Principles of Data Security

FOUNDATIONS OF COMPUTER SCIENCE

Series Editor: Raymond E. Miller

Georgia Institute of Technology

PRINCIPLES OF DATA SECURITY

Ernst L. Leiss

Principles of Data Security

Ernst L. Leiss
University of Houston
Houston, Texas

PLENUM PRESS • NEW YORK AND LONDON

Library of Congress Cataloging in Publication Data

Leiss, Ernst L., 1952–
Principles of data security.

(Foundations of computer science)
Bibliography: p.
Includes index.
1. Computers—Access control. 2. Data protection. 3. Electronic data processing departments—Security measures. I. Title. II. Series.
QA76.9.A25L48 1982 001.64′4 82-22272
ISBN 0-306-41098-2

© 1982 Plenum Press, New York
A Division of Plenum Publishing Corporation
233 Spring Street, New York, N.Y. 10013

All rights reserved

No part of this book may be reproduced, stored in a retrieval system, or transmitted in any form or by any means, electronic, mechanical, photocopying, microfilming, recording, or otherwise, without written permission from the Publisher

Printed in the United States of America

QA
76.9
.A25
L48
1982

Science
001.644
L53

Dedicated in gratitude
to my parents and my wife

Preface

With the greatly increased use of electronic media for processing and storing data, data security became an important practical issue. This is especially true for the extensively shared and distributed systems which are more and more being accepted in commercial operations. Essentially, the problem is that of protecting data, including all the implications which this has to the end user as well as the systems or database designer. However, usually the term *data security* refers to protection by technical, i.e., computer science specific, means; if one wants to include issues such as physical security, how to select the group of people who should have authority to perform certain operations, etc., the term *computer security* is more appropriate.

The object of this book is to provide technical solutions to (facets of) the problem of achieving data security. The reader who hopes to find clever recipes which allow circumventing protection mechanisms will, however, be sadly disappointed. In fact, we deliberately kept the presentation of the material at a fairly general level. While the short-term benefit of such an approach may be somewhat smaller, we feel that without a thorough understanding of the fundamental issues and problems in data security there will never be secure systems. True, people probably always considered certain security aspects when designing a system. However, an integrated approach is absolutely imperative. Furthermore, we believe that even persons who are not completely convinced of this thesis will eventually profit. The analogy with software development is quite apt. Software has been developed for over three decades; but only theoretical advances and design methodologies resulted in a significant improvement of the quality of software, the benefits of which we are only beginning to see today. Note that this applies even to software written by people or teams who do *not* strictly adhere to those methodologies. We expect the same phenomenon with data security. From this conviction stems our belief that a good deal of the seemingly theoretical material in this text is in fact very practical.

This book is addressed to more than one type of reader. In the university or teaching setting, it is appropriate for a graduate course as well as an advanced senior course in computer science or management information systems. In the industrial setting, it should be of interest to professionals who are dealing extensively with data which very frequently are proprietary, such are systems analysts and systems or database designers and managers. Finally, the book should also appeal to researchers who want to gain a comprehensive perspective of the field.

An effort was made to keep this book completely self-contained. We think we succeeded, except perhaps for one or two areas, where references are provided for readers who would like to get more information than is needed for the purpose of this book. We do, however, assume the usual level of mathematical maturity one might expect from a person with a Bachelor's degree in the sciences or in engineering. The material presented is based on lectures, namely, a summer course for advanced graduate students at the Universidad de Chile, Santiago, taught in 1979, and a regular graduate course at the University of Houston, taught in 1980 as well as in 1981.

The book is organized into three major parts, which are preceded by an introduction. The first part deals with the security of statistical databases, the second part discusses authorization systems, and the third part covers the fundamental ideas of cryptography. Despite the general presentation, heavy emphasis is on methods which are practically useful. The text contains a number of (not too difficult) exercises which form an integral part of the material; thus the reader should at least attempt a number of them. The book also contains a fairly extensive bibliography.

Ernst L. Leiss

Contents

Chapter 1

Data Security: Introduction and Motivation

In the last few years, the amount of information stored in electronic media has increased significantly. Information which is stored in some medium is usually called *data*. An important aspect of the storage of information is the way in which access is performed; this is commonly called *information retrieval*. Here, practically always some knowledge is required in order to be able to access the information. This knowledge can come in many different forms. For example, in order to be able to withdraw money from an automated teller, a certain code must be supplied. This code is assigned to the account holder personally and is to be kept confidential. It should be noted that this is indeed an example of limiting access to information, as it is the stored information ("How much money is left in the account?") which is ultimately translated into more tangible resources (i.e., bank notes). Of crucial importance is the requirement that without this knowledge the stored information cannot be accessed. Clearly, it would be highly undesirable if someone other than the account holder, i.e., someone who does *not* know the code, could withdraw money from the account. Since this vital knowledge is deliberately hidden, presumably by the owner of the resource, access becomes impossible for anybody but the original owner. Thus the resource is protected, the information is secret. Clearly, the owner can pass on this knowledge to selected persons who then also have access. For example, the code which enables one to withdraw money from an automated teller can be given to other persons; this act will confer the authority to withdraw money from a specific account. However, sometimes someone not in such a group is able to acquire the necessary knowledge in one way or another. Obviously this is undesirable and should be avoided if at all possible. This problem leads naturally to the question how well protected, in other words, how *secure*, the resource is.

Security is a measure of the *effectiveness* with which the resource is protected. For the purpose of this book it describes how difficult it is for an unauthorized person to obtain the knowledge required to access the confidential data. For example, if an account holder's code for the automated teller is easily obtainable (e.g., if it is printed on the checks), access to the account is virtually not protected, the resource is clearly not secure.

The desire to maintain certain information secret is nothing new; indeed it is probably as old as any method to store information. The fundamental question involved is the right to access data (and, through the data, resources). However, there is an important difference between physical resources and information, although access to physical resources *is* very often governed by access to information. A thing, say a car, is of use only to that person who possesses it; it exists uniquely. Information, on the other hand, can be used by anybody who knows, i.e., has access to it; it can be duplicated without losing its usefulness. Consider for example the exchange of funds between financial institutions. In earlier days, a physical object, money or gold, was exchanged. This object could be stolen but not duplicated. (We disregard the forging of bank notes, as forging is *not* perfect duplication.) Nowadays, this exchange of funds is achieved by exchanging *information*. It is conceivable that this information is intercepted and slightly modified (for example, the recipient could be changed). Then the modification as well as the original information are transmitted. The delay introduced into this transmission could be imperceptible to the receiving institution; furthermore both messages will appear genuine. This example demonstrates that additional safeguards are necessary to prevent the duplication of information which may allow unauthorized access to resources (in this case, funds). Thus, while it is sufficient to possess a thing to prevent anybody from using it, the situation in the case of information is entirely different; it must be kept secret to achieve this objective and it should be as secure as possible in order to maintain this inaccessibility to unauthorized persons.

While the desire for secrecy and security of data is quite old, recent developments in data processing and information retrieval mandate a reevaluation of existing methods for achieving these objectives. Only on the surface is this a question of quantity rather than quality. In fact, this (previously unheard of) amount of information does create *new* problems. To see this let us use an example which admittedly somewhat dramatizes the situation but nevertheless correctly illustrates the important point. The basic quality of a weapon is that it destroys. Thus the difference between a knife and a hydrogen bomb is only one of quantity: The bomb destroys more. However, in a wider context there are qualitative differences which are self-evident. A similar situation exists in the area of data security. Before the advent of high-speed computers the amount of data stored was rather limited, the information was stored in one place only, access was possible only to persons who were physically at the same place, it was laborious and time consuming, and finally it usually was possible in one way only (e.g., an office in a village which registered births and deaths when asked for the names of the men who are fathers of more than three but less than eight children who are still alive would be forced to sequentially search all the files several times, clearly a prohibitively time-consuming process). As a consequence, less information was asked to be supplied (because less could be handled) and security was much larger (as the information was highly distributed, retrieval required the physical presence of the person who wanted to obtain the information, and many queries were simply impossible or infeasible). A third group of observations pertaining to data security before the advent of computers can be summarized by the terms *smaller administrative units with little interaction* and *less mobility*; in other words, as people were born, worked, married, had children, and died more or less in the same place there was little need for an established way of information exchange. Furthermore the little exchange which did take place usually took on a very personal form.

Most of these (data-security-enhancing) restrictions on the interchange of information have been removed: Huge amounts of information are requested and stored. Often its relevance can only be guessed at, but the suspicion remains that at some point it could be used against its supplier. Access is almost unlimited; from virtually anywhere in the world almost anybody can pose a practically unlimited number and variety of queries in order to retrieve information from a vast diversity of sources. All that is required is a computer terminal as well as a telephone, in addition to the necessary authorization to use these resources (passwords, keys, etc.). Oftentimes, the same information is stored in more than one place, which, among other problems, creates the one that correction of wrong information in one place does not guarantee correction of all instances of this incorrect information (credit reports!). Interchange of information is all pervasive, mandated by the ever-increasing size of administrative units and by the mobility of people.

The objective of this book is limited to a discussion of technical solutions to problems inherent in data security. The reader should, however, be aware that many problems require other approaches (such as legislation). Our scenario is the following. Suppose we are given a collection of confidential data which we want to keep secret. However, the nature of the data requires that certain restricted access be granted, restricted either in terms of the group of persons who are authorized to use the data or in the sense that only information based on certain aspects of the data be divulged but not individual data items themselves. We are interested in a precise description of the kind of access which can be allowed without violating the confidentiality of the information. Our point of view is that of an end user in a database–data communication system; in particular, we will not worry about the external, i.e., physical, security of our system. Similarly, we will assume that the operating system performs as it is supposed to perform. Specifically, we will assume that the operating system is secure, i.e., does not contain features which allow circumvention of protection mechanisms. In view of recent results on secure operating systems this is not an entirely unreasonable assumption. (Readers who consider this assumption too controversial and unrealistic are reminded that only by verifying the correctness of one module at a time, the correctness of a large system can be demonstated.) Below follows a very brief summary of the material in the next three chapters including a justification of its selection at the expense of the exclusion of other topics.

We distinguish three main areas where security is of great importance and which are of relevance to a user of a given system. The first area is that of databases, in particular, statistical databases, where a user might be interested either in the question whether a given database with given access mechanism is secure (i.e., "Do I want to use this database to make my confidential collection of data available on a restricted basis?") or in the problem of designing his-her own database which satisfies certain requirements concerning its security under certain access mechanisms (i.e., "What kind of database do I have to design if I want this and that?"). The important observation here is that everybody should probably be allowed to extract information based on certain aspects of the data but should not be able to determine any particular data item.

The second area is that of authorization. The scenario here is as follows. A user has a data object which is to be accessible only to a restricted group of people, i.e., each of these persons is authorized to access it in a certain way. The kind of authorizations and the way in which they are processed is almost always beyond the user's influence. However, the owner of such a confidential data object is probably very interested in the question of whether a certain authorization which s/he intends to grant has an adverse effect on the overall security (i.e., "If I give A this right and B that right will that enable C to do something terrible to me?").

The third area, finally, is that of encryption. It can be considered another aspect of the previous problem. If the owner of a confidential data object allows a certain restricted group to access this object, secret information is transmitted. These transmitted messages could be intercepted by an outsider; thus the secrecy of the data object would not be maintained. A well-known scheme designed to avoid this is to encode the information which is to be transmitted in such a way that only the intended receiver is able to decode it, i.e., while (physical) interception is still possible, the intercepted message is useless as it cannot be understood by the intercepter (i.e., "Can I send you this confidential message without somebody else eavesdropping?").

All three areas we propose to investigate have in common that they are of very practical importance to anybody who owns confidential information which is used (obviously, if it is not used it can simply be locked up, in which case there are no problems about data security), and that to a large extent they are under the user's control. The chapter on authorization mechanisms (Chapter 3) does deal extensively with a problem which is of importance in operating systems, namely, the safety problem; however, we present it in a somewhat more general context and without making explicit reference to operating systems. Finally, another very important reason for choosing these three areas is the fact that within the last few years a considerable body of knowledge related to them has been accumulated and that in all three areas there are now practical and useful methods available to resolve the problems addressed here in a realistic way.

Bibliographic Note

A discussion of the questions of statistical database security, authorization mechanisms, and en- and decryption can be found in Denning and Denning[23] as well as a survey of other problems outside of the scope of this book, such as design of secure operation systems. This paper also contains an extensive bibliography.

Chapter 2

Statistical Database Security

2.1. Introduction

2.1.1. Motivation

A statistical database is a collection of data items each of which is to be considered confidential but access to statistical information about these data items is to be allowed. The central questions of this chapter can be formulated as follows: Is a given statistical database secure, i.e., using the responses to permitted queries is it possible to determine the previously unknown value of any of the confidential data items stored in the database, and more fundamentally, are there secure statistical databases?

At first this question appears strange. Since we do not allow access to individual data elements, it is intuitively clear that the confidentiality of our data is maintained, despite the fact that access to statistics is permitted. However, it turns out that our intuition is misleading. In this section we try to convince the reader informally that this intuitive feeling is by no means correct, in fact that there is a very definite problem in protecting the security of even the most restrictive statistical database. In subsequent sections we will give formal verifications of our assertions.

Consider a database containing grades of students who took a certain course, say, freshman composition. Each record has five keys, namely, SEX (N for female, Y for male), MEMBER of a fraternity/sorority (N for no, Y for yes), CLASS (from 1940 to 1979), OWNER of a car (N for no, Y for yes), and WORKING (N for no, Y for yes). We assume that each possible sequence of five appropriate symbols identifies at most one student; however, given such a sequence there need not exist a student satisfying the description. Each record has two other fields, one containing the NAME of the students, the other his/her GRADE (between 0 and 100). For the purpose of this example, the grade will be considered confidential information. Now suppose that the owner of this database allows a sociologist to obtain statistical information based on the data stored in the database. The security of the database is to be enforced in such a way that only queries which involve at least two entries of the database are permitted, in which case the response is the sum of the grades of all students involved in the query; otherwise the response is "not permitted query." For example, the sociologist might pose the query (*, N, 1977, Y, Y), i.e., the sum of the grades of all the students who were not in a fraternity or sorority, took the course in 1977, owned a car, and worked. The asterisk indicates that the students can be of either sex. If there are two students the response will be the sum of their grades; if there are fewer than two the response will be "not permitted query." Note that there cannot be more than two, one female and one male, but there could well be none. Our sociologist, who supposedly is unable to access individual records, now poses the following four queries and receives the corresponding responses:

q_1 (*, Y, 1978, N, Y) : 142

q_2 (*, Y, 1978, *, Y) : 206

q_3 (Y, Y, 1978, *, Y) : 134

q_4 (*, Y, 1978, Y, Y) : "not permitted query"

We claim that these four queries suffice to determine several individual grades. First we see that the record (N, Y, 1978, Y, Y) does not exist because (Y, Y, 1978, *, Y) is permitted, thus (Y, Y, 1978, N, Y) and (Y, Y, 1978, Y, Y) must exist but (*, Y, 1978, Y, Y) is not permitted. For brevity denote

(N, Y, 1978, N, Y) by x_1

(Y, Y, 1978, N, Y) by x_2

(Y, Y, 1978, Y, Y) by x_3

All three records exist. We now can rewrite the queries q_1, q_2, q_3 as a system of linear equations:

$$
\begin{aligned}
x_1 + x_2 \qquad &= 142 \qquad \text{(from } q_1\text{)} \\
x_2 + x_3 &= 134 \qquad \text{(from } q_3\text{)} \\
x_1 + x_2 + x_3 &= 206 \qquad \text{(from } q_2\text{)}
\end{aligned}
$$

and this is equivalent to

$$
\begin{bmatrix} 1 & 1 & 0 \\ 0 & 1 & 1 \\ 1 & 1 & 1 \end{bmatrix} \cdot \begin{bmatrix} x_1 \\ x_2 \\ x_3 \end{bmatrix} = \begin{bmatrix} 142 \\ 134 \\ 206 \end{bmatrix}
$$

Since this matrix is nonsingular we can solve the equations, and this yields

$$
\begin{aligned}
x_1 &= 72 \\
x_2 &= 70 \\
x_3 &= 64
\end{aligned}
$$

Hence we know that the students identified by (N, Y, 1978, N, Y) obtained 72, (Y, Y, 1978, N, Y) obtained 70, and (Y, Y, 1978, Y, Y) 64. The reader should realize that nothing was known beforehand about the database other than that an existing key sequence determines a unique element. Furthermore the intuitively very strong requirement that queries involving less than two elements will not be answered at all fails miserably and with it our intuition. In order to convince the reader that we did not cheat we reproduce the database for 1977 and 1978 in Table 2.1.

Table 2.1. A Simple Database

NAME	SEX	MEMBER	CLASS	OWNER	WORK	GRADE
A	N	Y	77	Y	N	74
B	Y	Y	78	N	Y	70
C	N	N	77	N	Y	86
D	Y	N	77	Y	Y	78
E	N	N	78	Y	N	67
F	Y	Y	78	Y	Y	64
G	N	Y	78	N	Y	72
H	Y	N	77	Y	N	69
I	N	N	77	Y	Y	73
J	Y	Y	77	Y	Y	60

While it is clear that this example demonstrates the failure of our intuition, it is not clear *why* it works. In order to provide insight into this "why" we will have to revert to a more formal presentation of the question of data security. We do hope, however, that this simple example provides sufficient justification for the more and more increasing concern about security of statistical databases as schemes similar to this have been (and are still being) used extensively in "real-world" databases.

2.1.2. Models and Methodology

In this section we first describe two models of statistical databases which are distinguished by the form of their queries. Both are simplifications of real-world databases but reflect the main properties which are important from the point of view of security. Then we define formally what we mean by the security of such databases.

First we introduce the characteristic-specified model (DC). A database in this case is a partial function DC from keys of length K into the real numbers R, where K is fixed throughout and at least 2, i.e.,

$$\mathrm{DC} : \{0, 1\}^K \to R$$

For instance, for $K = 2$, $\{0, 1\}^2 = \{00, 01, 10, 11\}$ and DC assigns to some or all elements a real number. If DC is a total function then $\mathrm{DC}(w)$ exists for all keys w in $\{0, 1\}^K$ otherwise there are some keys v for which $\mathrm{DC}(v)$ is not defined. The range of DC is the confidential information of the database we will be concerned with. Characteristic-specified queries of type f are strings from $\{0, 1, *\}^K$ with the following interpretation:

(a) f is a function of arbitrarily many arguments (e.g., average, median, maximum, minimum, etc.).

(b) If the database manager is presented with a query q in $\{0, 1, *\}^K$ of type f it first determines all keys w in $\{0, 1\}^K$ which are matched by q and then applies f to the values $\mathrm{DC}(w)$.

A key w is matched by a query q if for all $i = 1, \ldots, K$ either the ith positions of w and q are identical or the ith position of q is an asterisk (*). Since the information in such a database can obviously not be kept confidential if queries involving only one element are allowed we will require that a query return a value only if it matches at least two existing keys. (Generalizations to more than two keys are obvious.) In particular, if DC is a total function this is the same as saying that q in $\{0, 1, *\}^K$ must contain at least one asterisk. Clearly for partial functions this is not sufficient. This is due to the fact that the number of keys matched by a query q can normally not be precisely determined given the query alone; it can only be bounded from below by zero (as there may not be any key matched) and from above by 2^s, where s is the number of asterisks in q. In Section 2.2.1 we will do a case study of this model for the type f being summation.

We give two examples.

Example. The example in Section 2.1.1 can be rewritten in order to conform to this notation:

NAME	SEX	MEMBER	CLASS	OWNER	WORK	GRADE
A	0	1	0	1	0	74
B	1	1	1	0	1	70
C	0	0	0	0	1	86
D	1	0	0	1	1	78
E	0	0	1	1	0	67
F	1	1	1	1	1	64
G	0	1	1	0	1	72
H	1	0	0	1	0	69
I	0	0	0	1	1	73
J	1	1	0	1	1	60

The four queries q_1, q_2, q_3, and q_4 which compromise this database are then

$$*1101, \quad *11*1, \quad 111*1, \quad *1111$$

Example. Suppose we are given the following database:

I	II	III	IV
4	FR	1982	0.0
3	GR	1979	3.9
5	JR	1981	2.3
6	SO	1981	3.6
4	PB	1982	3.0
5	SR	1982	2.5
6	SO	1980	3.3
6	JR	1978	2.0

where column I gives the number of courses taken by this student in the present term, column II gives the student's standing, i.e., freshman (FR), sophomore (SO), junior (JR), senior (SR), postbaccalaureate (PB), or graduate (GR), and column III gives the year the student enrolled in the program, while column IV gives the student's gradepoint average (GPA). This can be reformulated as

I	II	III	IV
0100	001	01110	0.0
0011	110	01011	3.9
0101	011	01101	2.3
0110	010	01101	3.6
0100	101	01110	3.0
0101	100	01110	2.5
0110	010	01100	3.3
0110	011	01010	2.0

This allows students to take up to 15 courses, permits up to eight categories for the standing of a student, and covers the years 1969–1999 (plus 00000 for the years before 1969).

Sample queries might be the average GPA for all students whose enrollment started in 1981 or the median GPA of all students taking six courses. The first query would be transcribed as

*******01101 (answer here 2.95)

and the second query would result in

0110******** (answer 3.3)

The second model we will consider (in Section 2.3) is the key-specified one (DK). In this case a database is a total function

$$\mathrm{DK} : \{1, \ldots, N\} \rightarrow R$$

where N is the number of items in the database. Thus if $1 \leq i \leq N$, DK(i) is a real number, the result of query (i). DK(i) is the confidential information and i is (an encoding of) the key. Obviously queries (i) are not acceptable as they determine precisely one data item. Thus a key-specified k-query of type f where k is fixed and at least 2 is a sequence of k indices

$$(i_1, \ldots, i_k)$$

such that the following is true:

(a) f is a function of k arguments.

(b) Given the query $(i_1, \ldots, i_k)$ of type f, its result is

$$f(\mathrm{DK}(i_1), \ldots, \mathrm{DK}(i_k))$$

(c) A query $(i_1, \ldots, i_k)$ is permitted only if $i_j \neq i_m$ for $j \neq m$; otherwise $(i, \ldots, i)$ would be an acceptable query and for most of the commonly used types f (averages, medians, maxima, minima, etc.) this would be equivalent to direct access which we do not permit.

Example. Recall the first example on p. 11. If we consider only the column NAME as primary key then we can ask the following questions letting $k = 5$:

(a) Out of the students A, C, E, F, and J, how many are male?

(b) What is the median grade of A, B, C, D, and E?

In the first case we multiply the average over SEX for

$$(A, C, E, F, J)$$

by k to get the answer 2. In the second case we form the median over GRADE for

$$(A, B, C, D, F)$$

which turns out to be 74. It appears that the latter model (key-specified) is more general as one can pose direct questions (i.e., questions concerning specific entries) instead of being required to simulate this via qualifications and characterizations which only when combined with each other give the desired result. However, this impression is somewhat misleading as certain queries may not be permitted in the latter model while they are perfectly acceptable in the former. For example, while the question for the average grade of all male students can be formulated in both models [in the former it is 1**** with average over GRADE and in the latter it is the query returning the average over GRADE for (B, D, F, H, J)], the same query for all students who work can only posed in the first model (****1) since in the second model this would amount to posing the query (B, C, D, F, G, I, J), which clearly is not possible if $k = 5$ is assumed since there are seven indices present.

We now proceed to define security. Intuitively we want to say that the value of no element can be inferred from the responses to some sequence of permitted queries. However, typically a user does know some of the data in a database, namely, at least those entries which pertain to that user. For example, a student who took freshman composition in 1977 knows what grade s/he got. Thus we have the following definition:

Let D_0 be the set of keys for which the user knows the value in the database (D_0 is a subset of $\{0, 1\}^K$ in the characteristic-specified model and a subset of $\{1, \ldots, N\}$ in the key-specified model). Let Q_m be a sequence of m queries. If we can infer the previously unknown value DC(x) or DK(x) for some key x from the answer to the queries in Q_m then we say the database is compromised by Q_m (for DC, x is in $\{0, 1\}^K - D_0$; for DK, x is in $\{1, \ldots, N\} - D_0$). If for all $m \geq 1$ the database cannot be compromised by any sequence of m queries we say the database is secure.

This formal definition is quite rigid; in order to make it more applicable to practical, real-world situations we will relativize it to some extent.

(a) *Relative security.* If a given database is not secure but the smallest number of queries required to compromise is very large we will say that the database is relatively secure. In practical applications one must consider the cost of compromise and the value the confidential information has for the compromiser. The less important the information, the lower a requirement of security is acceptable. This kind of situation (relative security) does actually occur in secure databases when "too much" information is known to the malicious user (see Section 2.3.2.1).

(b) *Selective compromise.* In most practical cases one is not interested in accessing an arbitrary data item but a very specific one. Intuitively selective compromise is much stronger than simple compromise. This is basically correct; however, in many cases it turns out that selective compromise is just a little harder than simple compromise but definitely possible (e.g., for queries of type average or sum). Nevertheless there do exist cases where only simple compromise but not selective compromise is possible (see Section 2.3.1.3).

Consequently it makes sense to distinguish. Thus:

(c) *Partial compromise.* This means that part of the database cannot be accessed.

(d) *Global compromise*. In this case any element can be determined individually.

Mostly for compromisable databases we first look at compromise and then deduce selective compromise. For medians there is a marked difference in the complexities of simple and of selective compromise; for averages there is none. Furthermore, medians allow only partial compromise, while averages permit global selective compromise.

The two models are quite similar; nevertheless the characteristic-specified one is closer to reality than the key-specified one. It is not obvious how they compare to each other. The latter model is more extensively investigated than the former one. Fortunately, the positive results about secure databases to be presented in Section 2.3.2 using the key-specified model can be adapted to the characteristic-specified one without undue effort. Therefore the use of the key-specified model is sufficiently justified. First, however, we will show in the next section what is known about characteristic-specified queries.

Questions and Exercises

The following questions are intended to provide the reader with motivation and food for thought. At this point some of them may appear quite hard. Later sections will give precise methods for answering them; the reader is encouraged to come back to these questions and compare the answers with those derived from the later methods.

1. In this exercise we assume the database given in Table 2.1. Furthermore let the queries be characteristic specified. Suppose we want to determine the grade of D, i.e., GRADE(10011).

   (a) Assuming that you know already the grades of B, F, and G, show that two additional queries suffice for determining GRADE (10011).

   (b) Assuming that no grades are known to begin with, show that five queries suffice for determining GRADE(10011).

2. Extend question 1. to global compromise of the database given in Table 1.

3. Assume the characteristic-specified model, i.e., queries are elements of $\{0, 1, *\}^K$, for $K \geq 1$. Derive a formula for the number of all distinct queries. Note that this formula should depend only on K.

4. An s-query is a query in $\{0, 1, *\}^K$ with precisely s asterisks. How many such distinct s-queries are there? Note that this formula should depend only on K and s.

5. Can you compromise the database on page 12 using queries of type AVERAGE? Hint—Consider the following queries:

   0**1*1*01**1 : 3.10

   0*1**1001**1 : 3.75

   01**01*01101 : 2.95

6. Can you compromise the following database

   (a) with queries of type AVERAGE?
   (b) with queries of type MEDIAN?

w	DC(w)
0000	1
1000	2
1100	3
0100	4
0110	5
0111	6
0101	7

7. Consider the database of the first example on p. 11. Assume a query q in $\{0, 1, *\}^5$ returns the sum of the key GRADE provided there are at least *three* elements matched by q. For example *1*1* (which matches A, F, and J) is permitted, but 01*1* is not as it matches only A and G. Assuming that you know the complete database (except of course the column GRADE!) can you compromise the database? (Hint—Consider the queries 1***1 and 1**11!)

8. Can you extend the method indicated in 7 above to achieve global compromise of the database?

9. Consider a database

$$\mathrm{DK} : \{1, \ldots, 1000\} \rightarrow R$$

with key-specified queries.

(a) Let $k = 4$. Suppose the following queries q_i of type SUM give responses r_i:

q_i	r_i
(3, 9, 27, 81)	100
(3, 9, 27, 243)	180
(3, 9, 81, 243)	60
(3, 27, 81, 243)	40
(9, 27, 81, 243)	20

Can you determine any of the elements DK(3), DK(9), DK(27), DK(81), or DK(243)? How about all of them?

(b) Let $k = 3$. Suppose the following queries q_i of type MEDIAN give responses r_i:

q_i	r_i
(2, 5, 9)	1
(2, 5, 11)	1
(2, 9, 11)	1
(5, 9, 11)	1

Can you infer anything from these responses? What can you conclude if it is known that all elements in the database are pairwise distinct [i.e., $\mathrm{DK}(i) = \mathrm{DK}(j)$ implies $i = j$]?

(c) Let $k = 3$ again. Suppose the following queries q_i of type MINIMUM give responses r_i:

q_i	r_i
(2, 5, 9)	0
(2, 5, 11)	0
(2, 9, 11)	0
(5, 9, 11)	1

Can you infer anything from these responses? What can you conclude if it is known that all elements in the database are pairwise distinct?

BIBLIOGRAPHIC NOTES

The relational database model originated with Codd.[11–15] A commercial implementation is described in Astrahan *et al.*;[2] this paper also contains an extensive bibliography on this subject. The characteristic-specified model is taken from Kam and Ullman[48] and from Chin,[10] the key-specified model is described in Dobkin *et al.*[28] The two models are compared in a number of papers, e.g., in Denning,[21,22] which also contain a list of older references concerning security of statistical databases.

2.2. Security with Characteristic-Specified Queries

In this section we discuss two approaches to compromising databases with characteristic-specified queries. Both have in common that some information is required in order to permit compromise. However, only so little information is needed and, moreover, this information can practically always be obtained such that the (unadulterated) characteristic-specified model is virtually useless for confidential databases.

2.2.1. Queries of Type Sum: A Case Study

Let DC be the given database, DC : $\{0, 1\}^K \to R$. DC is assumed to be a total function, i.e., every key in $\{0, 1\}^K$ occurs actually in the database. Assume D_0 is empty, i.e., no entries of the database are known. An s-query is a sequence in $\{0, 1, *\}^K$ which contains exactly s asterisks where $1 \leq s \leq K$ and s is fixed throughout this section. In the following we will consider only s-queries of type SUM, i.e., the response to a query q in $\{0, 1, *\}^K$ is

$$\sum_{\substack{q \text{ matches} \\ w \text{ in } \{0,1\}^K}} \mathrm{DC}(w)$$

This model of a database is an abstraction of another model which is a precise representation of the actual use of relational databases. This latter model can be described as follows. Each record consists of category fields and data fields. Without undue restriction of the generality of the model we assume that there is only one data field. This data field constitutes the information to be protected. A query in this model uses a characteristic formula c which is an arbitrary logical expression using as terms logical expressions involving the values in the category fields as well as constants, the connectives being the operators AND, OR, and NOT, as well as comparisons, etc. If c is such a formula the corresponding query is defined as follows:

$$\sum_{\substack{\text{key } w \\ \text{satisfies } c}} \mathrm{DC}(w)$$

It is clear that s-queries as defined above can be transformed into queries using characteristic formulae. Consequently, databases with characteristic formulae are at least as compromisable as databases with s-queries. It is not so obvious whether the converse holds.

Let c be a characteristic formula and let the possible values of the category fields be encoded in binary. It follows that c can be expressed equivalently by a formula c' which operates on single bits instead of entire values of the category fields. Now we can see that it is not difficult to concoct a characteristic formula which requires two or more s queries each of which would match exactly one key and is therefore not permitted. It follows that databases with characteristic formulae are strictly less secure than databases with s queries. Thus all the negative results concerning security in the model with s queries will also hold for the other model. Another indication of this are the results in the following section.

First we observe that we can restrict our attention to 1-queries due to the following:

Lemma 1. If a database can be compromised by s-queries for $s \geq 2$ then it can also be compromised by $(s-1)$-queries.

Proof. Each s-query can be expressed as a sum of two $(s-1)$-queries. □

Now we can state the following:

Theorem 1. Assume that no elements are known, $D_0 = \emptyset$. Then no such database can be compromised by s-queries of type sum.

Proof. We show this for $s = 1$. Assume the contrary, i.e., that there exists a sequence Q_m of m 1-queries which allows us to determine one element of the database. We now define a new database DC′ as follows:

$$\mathrm{DC}'(w) = \begin{cases} \mathrm{DC}(w) + 1 & \text{if } \mathrm{WT}(w) = 0 \\ \mathrm{DC}(w) - 1 & \text{if } \mathrm{WT}(w) = 1 \end{cases}$$

for all w in $\{0, 1\}^K$ where the weight $\mathrm{WT}(w)$ of a key w is the number of ones in w modulo 2. It is clear that any 1-query applied to DC′ will yield exactly the same result as the same query applied to DC as there are two keys involved one being of weight 1 and the other being of weight 0. Thus our original assumption that DC can be compromised leads to a contradiction; DC cannot be compromised. □

Example. Consider the database given by

Key w	DC(w)
000	1
001	2
010	3
011	4
100	5
101	6
110	7
111	8

According to Lemma 1 any s-query can be expressed in terms of 1-queries. For example, the query *1* will return

$$3 + 4 + 7 + 8 = 22$$

This is the same as the sum of the 1-queries *10 (returning $3 + 7 = 10$) and *11 (returning $4 + 8 = 12$); another possibility are the two 1-queries 01* and 11*. Altogether there are exactly twelve 1-queries, each given with its response below:

*00	(6)	0*0	(4)	00*	(3)
*01	(8)	0*1	(6)	01*	(7)
*10	(10)	1*0	(12)	10*	(11)
*11	(12)	1*1	(14)	11*	(15)

According to the proof of the theorem we change the database as follows:

w	DC(w)
000	$1 + 1 = 2$
001	$2 - 1 = 1$
010	$3 - 1 = 2$
011	$4 + 1 = 5$
100	$5 - 1 = 4$
101	$6 + 1 = 7$
110	$7 - 1 = 6$
111	$8 - 1 = 7$

Now let us look at the 12 1-queries and the *new* responses:

*00	(6)	0*0	(4)	00*	(3)
*01	(8)	0*1	(6)	01*	(7)
*10	(10)	1*0	(12)	10*	(11)
*11	(12)	1*1	(14)	11*	(15)

As it is clear, the responses are the *same* although we changed every single element of the database. Thus as any response is a sum of responses to 1-queries, comproimse is not possible.

This result would be quite satisfying if it were not for some serious flaws. The first one is that the range of DC must be unknown; in other words, it is assumed that the user does not even know the approximate range of the numbers stored in the database. The second is the assumption that none of the values in the database are known to the user. The third flaw finally is the assumption that for each key in $\{0, 1\}^K$ there exists a record in the database. Any of these problems is serious; the three of them together render the result completely useless from a practical point of view. Thus in an attempt to salvage the situation let us see what happens if we drop the most inconspicuous of these assumptions namely that the range of the database be unknown.

Theorem 2. Assume that no data element is known, $D_0 = \emptyset$. A database DC : $\{0, 1\}^K \rightarrow \{0, 1, \ldots, d - 1\}$ is compromisable by 1-queries if and only if there exist keys v and w such that either (1) both v

and w are of equal weight, $WT(v) = WT(w)$, and $DC(v) - DC(w) = d - 1$, or (2) $WT(v) = 1$ and $WT(w) = 0$ and either $DC(v) + DC(w) = 0$ or $DC(v) = DC(w) = 2(d - 1)$.

Proof. (a) Suppose (1) or (2) hold; we have to show that the database can be compromised by 1-queries. Define the Hamming distance $Hd(v, w)$ of v and w to be the number of positions in which v and w differ. We first prove by induction on the Hamming distance the following.

Claim: If $v \neq w$ then by linear operations on the responses to 1-queries one can compute $DC(v) + DC(w)$ if $Hd(v, w)$ is odd and $DC(v) - DC(w)$ if $Hd(v, w)$ is even.

Basis: If $Hd(v, w) = 1$ then there exists a 1-query which matches precisely v and w and nothing else. The response to this query is $DC(v) + DC(w)$.

Induction: Suppose $Hd(v, w) = h > 1$. As DC is a total function there exists a key u with $Hd(v, w) = 1$ and $Hd(v, u) = h - 1$. If h is even we can compute $DC(v) - DC(w)$ by subtracting $DC(v) + DC(u)$ and $DC(u) + DC(w)$, which can be computed by the induction hypothesis and the fact that $h - 1$ is odd. Similarly if h is odd we can compute $DC(v) - DC(u)$ and $DC(u) + DC(w)$ and summing up yields $DC(v) + DC(w)$. This proves the claim. Observe now that $Hd(v, w)$ is odd iff exactly one of v and w has weight 0. Therefore we can compute all the sums and differences given in (1) and (2) by using 1-queries only. (Note that the theorem says only "if such v and w exist then compromise follows"; it does not postulate that v and w be known!) Now if (1) holds we can deduce $DC(v) = d - 1$ and $DC(w) = 0$, if (2) holds we obtain $DC(v) = DC(w) = 0$ from $DC(v) + DC(w) = 0$ and $DC(v) = DC(w) = d - 1$ from $DC(v) + DC(w) = 2(d - 1)$.

(b) Suppose that neither (1) nor (2) hold; we have to show that compromise is not possible. First we assume that for all keys v, $DC(v)$ is neither 0 nor $d - 1$. In this case we can use the same argument as in the proof of Theorem 1 to show that the database cannot be compromised. Therefore we will assume that for at least one key v_0, $DC(v_0)$ in $\{0, d - 1\}$. Let us first settle the case $DC(v_0) = d - 1$. By the negation of (1) we have $DC(w) \neq 0$ for all w with $WT(w) = WT(v_0)$. Thus define the new database DC′ as follows:

$$DC'(v) = \begin{cases} DC(v) + 1 & \text{if } WT(v) \neq WT(v_0) \\ DC(v) - 1 & \text{if } WT(v) = WT(v_0) \end{cases}$$

It follows that for all v, $DC'(v)$ is in $\{0, 1, \ldots, d-1\}$; thus no compromise is possible. Now suppose that there is no v such that $DC(v) = d-1$. Let v_0 be a key with $DC(v_0) = 0$ [if there is none we have the already settled situation $DC(v) \neq 0, d-1$ for all keys v!]. By the negation of (2) there is no w with $WT(w) \neq WT(v_0)$ such that $DC(v_0) + DC(w) = 0$, i.e., for all keys w with weight different from that of v_0, $DC(w) \neq 0$. Thus define DC′ as follows:

$$DC'(v) = \begin{cases} DC(v) + 1 & \text{if } WT(v) = WT(v_0) \\ DC(v) - 1 & \text{if } WT(v) \neq WT(v_0) \end{cases}$$

This yields a contradiction as $0 \leq DC'(v) \leq d-1$ for all keys v. Again, compromise is impossible. □

The upshot of this theorem is that even an extremely weak assumption about the range of the database invalidates the result of Theorem 1; the database can be compromised. Note that it is sufficient to start with a guess for the range (provided this assumed range is never smaller than the actual range) and then test whether the database satisfies Theorem 2. Clearly its proof yields an algorithm to do this, i.e., we have the following:

Corollary 1. There exists an algorithm which decides in time proportional to 2^K whether a database with finite range can be compromised with 1-queries.

Furthermore once one knows that the database can be compromised it is very easy to do it, as stated in the following:

Corollary 2. If the database can be compromised then at most $K-1$ 1-queries suffice to determine any further element (global, selective compromise).

Example. Consider the database $DC : \{0, 1\}^3 \rightarrow \{0, \ldots, 10\}$, i.e., the keys are 000, 001, 010, 011, 100, 101, 110, 111. Since every sequence of queries is equivalent to a sequence of 1-queries we can restrict our attention to the set of all possible 1-queries, given below together with their responses:

00*	2	0*0	10	*00	4
01*	16	0*1	8	*01	12
10*	14	1*0	12	*10	18
11*	8	1*1	10	*11	6

Let us first test whether condition (1) holds. There are four keys of even weight, namely, 000, 011, 101, and 110, and four keys of odd weight, namely, 001, 010, 100, and 111. For example, let us show how to compute DC(000) − DC(101); the other computations follow similarly. The query

0

matches the keys 000 and 101 but also two additional keys, namely, 001 and 100. Thus we determine the response to the query 00*, which is 2, and subtract the response to *01, namely, 12. This gives −10 for DC(000) − DC(101); thus we can conclude that

$$DC(101) - DC(000) = 10$$

i.e., the first condition of the theorem is satisfied. Already on the basis of this finding we know that the database can be compromised as it necessarily follows from

$$0 \leq DC(w) \leq 10 \qquad \text{for all keys } w$$

that DC(101) = 10 and DC(000) = 0. From this the database can be globally compromised.

Let us now test whether condition (2) holds. We must compute $DC(v) + DC(w)$, where v ranges over the four keys of odd weight and w over those of even weight. Out of these 16 possibilities we will compute two, namely,

$$DC(000) + DC(111) \text{ and } DC(010) + DC(101)$$

The following chain of queries goes from 000 to 111 (there are others, too):

000 -(00*)- 001 -(*01)- 101 -(1*1)- 111

The responses to 00*, *01, 1*1 are 2, 12, 10, respectively. We compute 2 − (12 − 10) = 0. Therefore,

$$DC(000) + DC(111) = 0; \qquad \text{hence } DC(000) = DC(111) = 0$$

Similarly, for DC(010) + DC(101),

010 -(*10)- 110 -(1*0)- 100 -(10*)- 101

and we get 18 − (12 − 14) = 20. Consequently, DC(010) + DC(101)

$= 20$. This implies that our database contains at least two instances where condition (2) is met. Thus, according to this condition it can be compromised.

Let us now consider the second problem we had with Theorem 1, namely, the assumption that no value is known to the user. Suppose we drop this assumption and assume that several values are known, i.e., we know DC(v) for all v in V_0 which is a subset of $\{0, 1\}^K$.

Proposition 1. If one value of the database is known the entire database can be compromised using 1-queries (global, selective compromise).

Proof. Let DC(v_0) be known for some key v_0. Given an arbitrary key w we determine DC(w) by induction on Hd(v_0, w). If Hd(v_0, w) $= 1$ the claim follows trivially (pose that 1-query which matches just v_0 and w). If Hd(v_0, w) $= h > 1$ there exists a key v such that Hd(v_0, v) $= 1$ and Hd(v, w) $= h - 1$. Since we now can determine DC(v) by the induction hypothesis the claim follows. □

Thus even knowing just one data entry is sufficient to compromise the database. Note, however, that this is only for the case of 1-queries. It can be shown that for global compromise by s-queries $O(K^{s-1})$ data items must be known and that one can always choose the items in such a way that this is also sufficient.

Let us now drop the third assumption in Theorem 1, namely, that DC is a total function, i.e., that DC(w) exists for all keys w in $\{0, 1\}^K$. Since total functions are much easier to handle than partial functions we make our partial function DC into a total one as follows. Our database DC will be a total function from keys to "flagged" numbers,

$$\mathrm{DC} : \{0, 1\}^K \to \{0, 1\} \times R$$

and DC(w) $= (b_w, r_w)$ means that for $b_w = 1$ the result is the number r_w (the record exists) but $b_w = 0$ indicates that the record does not exist thus in this case r_w is irrelevant. In this way DC is still a total function but models accurately the original partial one. An s-query is defined as above, i.e., q is an element of $\{0, 1, *\}^K$ with exactly s asterisks but now it matches only those keys w for which the flag b_w is 1 (existing records). While before it was easy to distinguish permitted queries from those which were not—those with $s \geq 1$ were permitted and the others not—now we have to be more careful as it is possible that an s-query even for arbitrarily large s does not match sufficiently many keys. Thus we will postulate that only those s-queries (for $s \geq 1$ arbitrary) be permitted which involve at least two (existing) keys; otherwise the system will reply "not permitted query." Now we can state the following:

Theorem 3. If one knows that a particular key v_0 exists then one can determine the existence of all records in the database.

Proof. By induction on the Hamming distance $\mathrm{HD}(v_0, w)$. If $\mathrm{Hd}(v_0, w) = 1$ then there exists a 1-query matching exactly v_0 and w. If the response to this query is "not permitted query" we know that the key w does not exist; otherwise key w does exist. Let $\mathrm{Hd}(v_0, w) = h > 1$ and assume that the induction hypothesis is true for all keys with a Hamming distance less than h from v_0. Let q be that h-query which has an asterisk whenever v_0 and w differ. If the response to q is "not permitted query" we know that the record with key w does not exist. Otherwise there either exists a key v with $b_v = 1$ which is matched by q; this implies that $\mathrm{Hd}(v_0, v) < h$ and $\mathrm{Hd}(v, w) < h$ thus by the induction hypothesis we can determine b_w. Or there does not exist such a key v then b_w must be 1 as the query q was permitted. Note that the existence of such a key can be effectively determined. □

In a very similar way we can also show this for the values of a database; namely, we have the following:

Corollary 3. If the value of some record is known then the values of all the records in the database can be determined (global, selective compromise).

For the following we change the response to a query somewhat. Rather than just returning the sum of the values whose keys are matched (or "not permitted query") the response will contain additionally the number of existing keys which were matched by this query. More formally if q is a query the response to q is the pair

$$(\mathrm{NU}(q), \mathrm{SUM}(q))$$

or "not permitted query," where $\mathrm{NU}(q)$ is the number of keys matched by q and $\mathrm{SUM}(q)$ is the sum of the corresponding values. With each database DC we associate an undirected graph

$$G = (V, E)$$

called the query graph which has the set of all existing records as set of nodes,

$$V = \{w \mid b_w = 1\}$$

and $\{v, w\}$ is an edge in E if there exists a query q such that v and w are the only keys matched by q, $E = \{\{v, w\} \mid \text{there exists a query matching } v \text{ and } w \text{ only}\}$. Now we can characterize compromisable databases within the framework considered in this section.

Theorem 4. If we know the existence of a specific record, i.e., $b_w = 1$ for some key w then the database DC can be compromised globally and selectively iff (a) the query graph for DC has at least one cycle of odd length or (b) there exists a query p such that NU(p) is odd and at least 3.

Proof. The proof follows by a sequence of lemmas. Note that by Theorem 3 we know b_w for all keys w. By Corollary 3 all we have to determine is the value r_u for one (existing) key u.

Lemma 2. If b_w is known for all keys w then the database can be globally and selectively compromised if condition (a) holds.

Proof. By definition of the query graph of a database the existence of a cycle of odd length, say, n, yields a system of n linear equations of the form

$$\begin{aligned} r_1 + r_2 &= \mathrm{SUM}(q_1) \\ r_2 + r_3 &= \mathrm{SUM}(q_2) \\ r_3 + r_4 &= \mathrm{SUM}(q_3) \\ &\vdots \\ r_n + r_1 &= \mathrm{SUM}(q_n) \end{aligned}$$

or in matrix formulation

$$\begin{bmatrix} 1 & 1 & 0 & 0 & \cdots & 0 \\ 0 & 1 & 1 & 0 & \cdots & 0 \\ 0 & 0 & 1 & 1 & \cdots & 0 \\ & & & \vdots & & \\ 1 & 0 & 0 & 0 & \cdots & 1 \end{bmatrix} \cdot \begin{bmatrix} r_1 \\ r_2 \\ r_3 \\ \vdots \\ r_n \end{bmatrix} = \begin{bmatrix} \mathrm{SUM}(q_1) \\ \mathrm{SUM}(q_2) \\ \mathrm{SUM}(q_3) \\ \vdots \\ \mathrm{SUM}(q_n) \end{bmatrix}$$

It is easily verified that this matrix is nonsingular (note that n is odd,

$n \geq 3$); therefore the system can be solved, the database can be compromised. By Corollary 3 this compromise is global and selective. □

Lemma 3. If b_w is known for all keys w then the database can be globally and selectively compromised if condition (b) holds.

Proof. We first claim that for any query q with $\text{NU}(q) \geq 3$ there exist two queries q' and q'' such that

$$\text{NU}(q') < \text{NU}(q) \qquad \text{or} \qquad \text{NU}(q'') < \text{NU}(q)$$

and q', q'' are defined as follows: For one position where q has an asterisk (*), q' has a zero (0), and q'' has a one (1), while all other positions remain unchanged. First we observe that q' and q'' together match exactly the same keys which q matches and that no key is matched by q' and q'' at the same time; thus $\text{NU}(q') + \text{NU}(q'') = \text{NU}(q)$. Since $\text{NU}(q) \geq 3$ there exist three keys matched by q, consequently there exists one position where two of these three keys differ. Hence q must have an asterisk in this position. Let this be the position in which q' and q'' differ. It follows that both queries match strictly fewer keys than q. As $\text{NU}(q) \geq 3$ at least one of q' and q'' must be permitted (i.e., matches at least two keys) which proves the claim.

Now we proceed to prove the lemma by induction on $\text{NU}(q)$. If $\text{NU}(q) = 3$ one of $\text{NU}(q')$ and $\text{NU}(q'')$, say, $\text{NU}(q')$, is 2; hence $\text{NU}(q'')$ is not permitted. Since we know b_w for all keys w we can determine the only key, say, u, matched by q''. Consequently,

$$\text{DC}(u) = \text{SUM}(q) - \text{SUM}(q')$$

Now we assume that $\text{NU}(q)$ is odd and at least 5. Without loss of generality let $2 \leq \text{NU}(q') < \text{NU}(q)$. If q'' is not permitted we compromise as above. If q'' is permitted one of $\text{NU}(q')$ and $\text{NU}(q'')$ must be odd (and therefore at least 3). Since q' and q'' are permitted we have $\text{NU}(q') < \text{NU}(q)$ and $\text{NU}(q'') < \text{NU}(q)$. Thus we can apply the induction hypothesis to compromise the database. By Theorem 2 the compromise is global and selective. □

This concludes the proof that conditions (a) and (b) are sufficient for global and selective compromise. We now show that they are also necessary.

Lemma 4. If b_w is known for all keys then the database can be compromised only if either (a) or (b) hold.

Proof. Assume neither (a) nor (b) hold. First we claim that the response to any sequence of queries can be written as a system of linear equations each of which consists of exactly two terms. Consider a typical query SUM(q). As b_w is known for all keys w

$$\mathrm{SUM}(q) = \mathrm{DC}(v_1) + \cdots + \mathrm{DC}(v_m)$$

where $v_1, \ldots, v_m$ are all the existing keys matched by q. Since (b) does not hold m is always even. By the argument in the proof of Lemma 3 we can always find valid queries q' and q'' such that NU(q') and NU(q'') are even and less than NU(q) provided $m \geq 4$. Since SUM(q'') + SUM(q') = SUM(q) the linear equation for q is redundant. Consequently we can split the original query q into queries each consisting of exactly two keys. We eventually end up with a system of linear equations each of which has exactly two terms. Now we must prove that for no key w, DC(w) can be determined from such a system of linear equations. Let us construct another database DC′ as follows. Consider the query graph G of DC. Choose an arbitrary node v in G and change r_v by adding 1. For all nodes v' immediately adjacent to v change $r_{v'}$ by subtracting 1. In general for all nodes v' adjacent to a node v where r_v was changed by adding (subtracting) 1, change $r_{v'}$ by subtracting (adding) 1. This is possible as G has only cycles of even length for (a) does not hold. Therefore we changed every record in the database but the responses to the queries to which corresponds the above system of equations remain unchanged; note that each equation corresponds to an edge in the query graph. Therefore the database cannot be compromised.

This concludes the proof that the two conditions in the theorem are also necessary; therefore Theorem 4 is proven. □

Note that this theorem contains as a special case Theorem 1 as it is easily verified that for the kind of database considered there neither condition (a) nor condition (b) of Theorem 4 are satisfied. However, as already pointed out it is not likely that this occurs in practice. In order to avoid this unpleasant situation it has been suggested to limit the number of queries a user is allowed to pose. Therefore it is of importance to know how many queries are necessary to achieve compromise. From the proof of Theorem 3 the following upper bound is obtained.

Proposition 2. If r_{v_0} is known for some key v_0 then at most $2^j - 1$ queries are sufficient to determine r_w, where j is the Hamming distance between v_0 and w, $j = \mathrm{Hd}(v_0, w)$.

However, this upper bound is unrealistically large. In fact one can show that on the average $(1 + p^{-1})^{\log(K)}$ queries suffice, where p is a measure of the density of the database, namely, the ratio between the number of existing keys and the total number of possible keys, $0 < p \leq 1$. We will not give a proof of this result as in practice it is completely unfeasible to limit the number of queries a user is allowed to pose in order to enforce security of a database. We will discuss this problem later (see Section 2.3.1.2) in a wider context; suffice it here to mention the problem of two or more users combining their results.

2.2.1.1. Some Remarks on Query Graphs

The query graphs used in the previous section have some interesting properties, some of which are discussed below. It should be noted that the motivation for these query graphs comes from the restriction that a query be permitted only if it involves at least $h = 2$ elements of the database. Clearly, this could be generalized to $h \geq 2$; the graph theoretic formulation would then involve hypergraphs. However, no complete characterization analogous to Theorem 4 has been obtained; some (partial) results are mentioned in this section.

First we want to test whether a graph has only cycles of even length. This of course is the basis of the proof of Lemma 3: If a query graph does not have any cycles of odd length then we can change the database in such a way that the responses to all the 1-queries remain unchanged. We employ the following algorithm for the purpose of testing this.

Algorithm:

Input: An arbitrary undirected finite graph without multiple edges or self-loops.

Output: YES if the given graph contains only cycles of even length, NO if the graph contains at least one cycle of odd length.

Method: (i) Pick one node and assign + to it; mark this node.

(ii) **For** all unmarked nodes v with + or − assigned **do**
assign the sign opposite to v's to all nodes which are connected to v; mark v;
if in this way a node gets assigned a sign different from a previously assigned sign, return "NO" and stop.
od

(iii) Return "YES" and stop.

Note that this algorithm actually employs the method in which the database is changed in the proofs of Theorem 1 and Lemma 3. However, if this modifying of the database cannot be performed consistently, e.g., if at some point we are to subtract from an element to which we had to add, then we are certain that there is a cycle of odd length present in this graph.

Example. Consider the following graph:

$$G = (\{1, \ldots, 20\}, E)$$

with the set of edges E given by

$\{1, 2\}$ $\{1, 3\}$ $\{2, 4\}$ $\{2, 5\}$ $\{3, 4\}$ $\{3, 20\}$ $\{5, 6\}$
$\{5, 7\}$ $\{6, 8\}$ $\{7, 8\}$ $\{8, 9\}$ $\{9, 10\}$ $\{9, 11\}$ $\{10, 12\}$
$\{11, 12\}$ $\{12, 13\}$ $\{13, 14\}$ $\{13, 15\}$ $\{14, 16\}$ $\{15, 16\}$ $\{16, 17\}$
$\{17, 18\}$ $\{17, 19\}$ $\{18, 20\}$ $\{19, 20\}$

Performing this above indicated test yields the following:

1	|		+		+					
2		—		—						
3				—						
4			+		+					
5			+		+					
6				—		—				
7				—		—				
8					+		+			
9						—		—		
10							+		+	
11							+		+	
12								—		
13								—	+	*** not possible ***
14							+			
15							+			
16						—		—		
17					+		+			
18				—						
19				—		—				
20			+		+					

Thus node 13 would get the signs + *and* −, which is not possible. Consequently this graph has a cycle of odd length. Clearly the algorithm is of linear time complexity. This should be contrasted with the problem of finding a Hamiltonian cycle in an arbitrary graph; this question is well known to be NP complete (see, for instance, Karp[49]).

An interesting problem in this context is the question which graphs can occur as query graphs. First we summarize some basic implications of the definition of query graphs.

Fact. A query graph is a finite connected undirected graph without loops and without multiple edges.

Proof. The finiteness follows from the fact that there are only 2^K different strings of K zeros and ones for any K. The absence of loops, i.e., edges of the form $\{v, v\}$, and of multiple edges follows directly from the definition. Thus we have to show that any query graph is connected. Let us assume that the given query graph is not connected; consequently it consists of a finite number of connected components. Let v and w be vertices of minimal Hamming distance in two different components, and consider $q(v, w)$ (the minimal query matching v and w). As v and w are not connected (they are in different components!) $q(v, w)$ must match at least one other node, say, u. Clearly,

$$\mathrm{Hd}(v, u) < \mathrm{Hd}(v, w) \quad \text{and} \quad \mathrm{Hd}(w, u) < \mathrm{Hd}(v, w)$$

Thus u must be a third component as $\mathrm{Hd}(v, w)$ is minimal for their two components. Iterating this argument with u in place of w will eventually yield a contradiction as there is only a finite number of different components. □

From now on whenever we use the term *graph* in this section we mean a finite connected undirected graph without loops and without multiple edges. Next we show that line graphs, cycle graphs, and complete graphs allow direct representation as query graphs.

A line graph on n vertices ($n \geq 1$) is a graph $\mathrm{LG}_n = (\{v_1, \ldots, v_n\}, E_n)$, where $E_n = \{\{v_i, v_{i+1}\} \mid i = 1, \ldots, n-1\}$.

Fact. Every line graph LG_n can be represented as a query graph.

Proof. Let $V' = \{v_1', \ldots, v_n'\}$, where each v_v' is defined by

$$v_i' = 0^{n-i}\, 1^{i-1} \qquad \text{for } i = 1, \ldots, n$$

We claim the $E(V')$, the set of edges of the query graph determined by V', is exactly $E' = \{\{v_i', v_{i+1}'\} \mid i = 1, \ldots, n-1\}$. It is clear that E' is contained in $E(V')$ as $\{v_i', v_{i+1}'\}$ is matched by the query

$$0^{n-i-1} * 1^{i-1}$$

and no other vertex is matched by this query. Thus it remains to show that there are no other edges in $E(V')$. Assume that there is an edge $\{v_i', v_j'\}$ with $|i - j| > 1$. Without loss of generality let $i > j$. Then $q(v_i', v_j')$, the query defined by putting an asterisk wherever v_i' and v_j' differ, has precisely $\mathrm{Hd}(v_i', v_j') = i - j$ asterisks; more specifically,

$$q(v_i', v_j') = 0^{n-1} *^{i-j} 1^{j-1}$$

However, this query matches precisely $i - j + 1$ vertices; consequently the corresponding edge cannot be in $E(V')$. □

A cycle graph on n vertices is a graph $\mathrm{CG}_n = (\{v_0, \ldots, v_{n-1}\}, E_n)$, where $E_n = \{\{v_i, v_j\} \mid j = (i+1) \bmod(n) \text{ for } i = 0, \ldots, n-1\}$.

Fact. Every cycle graph CG_n can be represented as a query graph.

Proof. For $n = 1, 2$, CG_n is clearly LG_n. For $n = 3$, one verifies directly that the query graph determined by $V = \{000, 011, 110\}$ is isomorphic to CG_3. Therefore let $n \geq 4$. Define $t = (n+1)/2$ if n is odd and $t = n/2$ if n is even. Define for $i = 0, 1, \ldots, t-1$ and for $j = 1, \ldots, t-1$

$$v_i' = 0^{t-i}\, 1^i, \qquad v_{n-j}' = 1^{t-j}\, 0^j$$

and if n is even, also

$$v_t' - 1^t$$

Let V' be the set of these n keys. We claim the $E(V') = E'$, where $E' = \{\{v_i', v_j'\} \mid \{v_i, v_j\} \text{ in } E_n\}$. First we show the E' is contained in $E(V')$. Clearly, for $i = 0, 1, \ldots, t-2$ the keys v_i' and v_{i+1}' are precisely matched by

$$0^{t-i-1} * 1^i$$

while for $i = 1, \ldots, t-2$, the query

$$1^{t-i-1} * 0^i$$

matches precisely the keys v_{n-i}' and v_{n-i-1}'. Furthermore, v_0' and v_{n-1}'

are matched by

$$* \, 0^{t-1}$$

None of these queries can match more than two keys as they contain only one asterisk. If n is odd then the query $* \, 1^{t-2} \, *$ matches $0 \, 1^{t-1} = v'_{t-1}$ and $1^{t-1} \, 0 = v'_{n-t+1}$ and no other vertex. However, if n is even we also have the key v_t' and thus two other queries with one asterisk each, namely,

$$* \, 1^{t-1} \qquad \text{and} \qquad 1^{t-1} \, *$$

Consequently E' is contained in $E(V')$. For the converse we must show that no edges other than those in E' exist in $E(V')$. By the argument in the previous proof it is clear that no pair $\{v_i', v_j'\}$ for i, j in $\{0, 1, \ldots, t-1\}$ nor $\{v'_{n-i}, v'_{n-j}\}$ for i, j, in $\{1, \ldots, t-1\}$ can be connected by an edge in $E(V')$ if $|i-j| > 1$. If n is even this also applies to $\{v_t', v_i'\}$ for $i = 0, \ldots, t-2$ and to $\{v_t', v'_{n-i}\}$ for $i = 1, \ldots, t-2$. Thus all we have to show is that it is impossible to have an edge $\{v_i', v'_{n-j}\}$ in $E(V')$ for i in $\{0, \ldots, t-1\}$ and j in $\{1, \ldots, t-1\}$. If $i \leq j$ we have

$$q(v_i', v'_{n-j}) = *^{t-j} \, 1^{j-i} \, *^i$$

if $j < i$ then

$$q(v_i', v'_{n-j}) = *^{t-i} \, 0^{i-j} \, *^j$$

In each case it is clear that there are more than those two keys matched by the query. This concludes the proof. □

A complete graph on n vertices is a graph $K_n = (\{v_1, \ldots, v_n\}, E_n)$ with $E_n = \{\{v_i, v_j\} \mid i \neq j, \, i, j \text{ in } \{1, \ldots, n\}\}$.

Fact. Every complete graph K_n can be represented as a query graph.

Proof. The cases $n = 1, 2, 3$ are covered by the previous results. Therefore assume $n \geq 4$. Define for $i = 1, \ldots, n$

$$v_i' = 0^{i-1} \, 1 \, 0^{n-i}$$

and let $V' = \{v_i' \mid i = 1, \ldots, n\}$. We have to show $E(V') = E'$, where $E' = \{\{v_i', v_j'\} \mid i \neq j\}$. We claim that $E(V')$ consists of all queries composed of $n-2$ zeros and two asterisks. It is not difficult to see that any two of the vertices are matched by such a query and no other vertex is matched by this particular query. More specifically, consider v_i' and v_j'; the corresponding query has an asterisk in positions i and j, and has zeros everywhere else. As no other key possesses this pattern of zeros no other vertex is matched. There are precisely $n(n-1)/2$ different such queries, and there are exactly $n(n-1)/2$ edges in a complete graph K_n. This concludes the proof. □

We can draw an interesting conclusion from the last result: Every finite undirected graph without loops and without multiple edges can be embedded in a query graph.

Additional constructions are described in Leiss[56]; recently Peter M. Winkler[100] showed that any (finite connected undirected) graph (without self-loops and without multiple edges) can in fact be realized as a query graph. The proof uses the fact that in any such graph G there exists a spanning tree with certain properties which allows to construct directly a set V of keys such that G is isomorphic to the query graph induced by V.

We conclude this section with some remarks on generalizing the concept of a query graph. Recall that so far we permitted queries which matched at least two existing keys. It is natural to generalize this such that a query is permitted only if it matches at least h existing keys, for h fixed, $h \geq 2$. We first observe that compromise may still be very easy to attain regardless how large h is:

Lemma. Let q be a query matching precisely s keys, and let q be a query which contains q and matches precisely $s+1$ keys. Then the database can be compromised by just those two queries.

Proof. Let q match the keys $v_1, \ldots, v_s$, and let q' match the keys $v, v_1, \ldots, v_s$. Then clearly $\mathrm{DC}(v) = \mathrm{SUM}(q') - \mathrm{SUM}(q)$. □

Note, however, that not every database permits this. For example, if we have the keys 000, 001, 010, 011, 100, 101, 110, 111 then every query will match an even number of keys.

Let us define an h-query graph $G = (V, {}_hE)$ as follows: The vertices of G are the (existing) keys of the database. There will be an edge in ${}_hE$ connecting the vertices $v_1, \ldots, v_h$ iff there exists a query matching precisely those keys and no others. It should be obvious that h-query graphs need not be connected any longer for $h > 2$. For example, if we are given the set of keys

$$\{000, 001, 010, 011, 100, 101, 110, 111\}$$

then the 3-query graph for these keys has no edge at all, while the 4-query has six edges, listed below together with the corresponding queries:

0**	{000, 001, 010, 011}
1**	{100, 101, 110, 111}
0	{000, 001, 100, 101}
1	{010, 011, 110, 111}
**0	{000, 010, 100, 110}
**1	{001, 011, 101, 111}

Let us call an m-loop in an h-query graph $G = (V, {}_hE)$ a set of m keys $v_0, \ldots, v_{m-1}$ together with m queries $q_0, \ldots, q_{m-1}$ such that q_i matches precisely the vertices v_j with

$$j \text{ in } \{(i + s) \bmod(m) \mid s = 0, \ldots, h - 1\}$$

Then we can state the following:

Proposition. Suppose a query is permitted only if it matches at least h keys. Assume there is an m-loop in the h-query graph of the database. Then the queries of the m-loop compromise the database iff $m \neq 0 \bmod(h)$.

Proof. The condition of the proposition gives rise to a system of linear equations of the following form:

$$\begin{bmatrix} 111 & \cdots & 1100 & \cdots & 0 \\ 011 & \cdots & 1110 & \cdots & 0 \\ 001 & \cdots & 1111 & \cdots & 0 \\ & \ddots & & \ddots & \\ 111 & \cdots & 1000 & \cdots & 1 \end{bmatrix} \cdot \begin{bmatrix} DC(v_0) \\ DC(v_1) \\ DC(v_2) \\ \vdots \\ DC(v_{m-1}) \end{bmatrix} = \begin{bmatrix} SUM(q_0) \\ SUM(q_1) \\ SUM(q_2) \\ \vdots \\ SUM(q_{m-1}) \end{bmatrix}$$

where each row in the matrix (as well as each column) contains exactly h ones, or

$$M(m, h)^* D_m = Q_m$$

It follows that the determinant of $M(m, h)$ is $\pm h$ if $m \neq 0 \bmod(h)$ and it is 0 if $m = 0 \bmod(h)$. Consequently the system can be solved in the $DC(v_i)$ iff $m \neq 0 \bmod(h)$. □

We terminate this section by showing that for general h there are databases which can be compromised if queries are permitted which match at least h keys but cannot be compromised if queries must match at least $h + 1$ keys.

Proposition. Let a query be permitted only if it matches at least h keys. For any $s \geq 2$ there exist databases which can be compromised with $h = s$ but not with $h = s + 1$.

Proof. Let $m = 2s + 1$, and consider the cycle graph on m nodes. By a previous result the graph CG_m can be realized as query graph. There are exactly m queries matching precisely s keys; furthermore these queries constitute an m-loop in the s-query graph. As $m = 1 \bmod(h)$ the above proposition demonstrates that compromise is possible. However, if $h = s + 1$ one can see that there are only $s + 1$ permitted queries and these queries cannot be used to achieve compromise. □

Questions and Exercises

1. Show that Lemma 1 is not true anymore if we do not require DC to be total.

2. In the example on p. 23, compute

$$DC(v) + DC(w)$$

for $v = 010$ and $w = 000, 011, 101, 110$; also compute

$$DC(v) - DC(w)$$

for $v = 011$ and $w = 101, 110$, and for $v = 111$ and $w = 001, 100$.

3. Give a detailed algorithm for determining whether a database with finite range can be compromised (Corollary 1). Make sure that it requires only 2^K queries.

4. Give a detailed algorithm which determines any other database element in at most $K - 1$ 1-queries when given some key v together with $DC(v)$.

5. Give a detailed algorithm for determining the values of all 2^K database elements in precisely $2^K - 1$ 1-queries when given DC(v) for some key v. Note that you cannot simply call the algorithm of Exercise 4, for each unknown element; the computations must be suitably rescheduled and organized.

6. Show that the assumption that a query q return

$$(\mathrm{NU}(q), \mathrm{SUM}(q))$$

instead of just SUM(q) is not very restrictive. More specifically, show that the two models are equivalent if it is known that a particular key v_0 exists.

7. Consider the database

$$\mathrm{DC} : \{0, 1\}^3 \rightarrow \{0, \ldots, 7\}$$

where the value returned by any 1-query abc in $\{0, 1\}^3$ is defined as follows: $4a' + 2b' + c'$, where $0' = 0$, $1' = 2$, and $*' = 1$. For example, the response to query *11 is $4*' + 2 \cdot 1' + 1' = 4 + 4 + 2 = 10$. Note that the response to **1 is not defined as this is not a 1-query. Determine whether this database can be compromised. If yes determine DC(w) for all w in $\{0, 1\}^3$.

8. Find the query graph for $\{0, 1\}^K$ for $K = 2, 3, 4$. Compare Theorems 1 and 4 for these graphs.

9. Find the query graph for the Exercise 6, p. 16. Compare your response there with that given by Theorem 4.

10. Find the query graph for the first example on p. 11. Is the compromise on p. 8 the same which would result by applying Theorem 4? Determine all possible ways in which this database can be compromised.

11. Write a program which accepts as "input" a database where the response to a query is either the sum of the matched elements, or—if there are less than two matched keys—"not permitted query." Let v_0 be a key which is known to exist (also input to the program). Write a subroutine which determines which keys exist (i.e., imple-

ment Theorem 3). Write the subroutine in such a way that it can also be used to implement Corollary 3.

12. Can you find query graphs for the following graphs: $G = (V, E)$ with

    (a) $V = \{v_0, \ldots, v_8\}$
    $E = \{\{v_i, v_{i+1}\} \mid i = 0, \ldots, 7\} \cup \{\{v_0, v_4\}, \{v_0, v_5\}, \{v_0, v_8\}\}$

    (b) $V = \{v_0, \ldots, v_6\}$
    $E = \{\{v_0, v_i\} \mid i = 1, \ldots, 6\}$.

13. Can you generalize 12(b), i.e., can you give query graphs for the following class of graphs $G = (V_n, E_n)$ where

$$V_n = \{v_1, \ldots, v_n\} \qquad \text{and} \qquad E_n = \{\{v_1, v_i\} \mid i = 2, \ldots, n\}$$

14. Determine the determinant of the matrix $M(m, h)$.

15. Construct the h-query graph for $h = 2, 3, 4, 5, 6$ for the following sets of keys (vertices):

    (a) {0000, 0010, 1011, 1111, 0110, 0001}

    (b) {011111, 101111, 110111, 111011, 111101, 111110}

    Does any of these h-query graphs have an m-loop which achieves compromise of the corresponding database?

16. Prove the correctness of the algorithm on p. 30.

17. Prove the following assertions used in the proof of Theorem 2:

    (a) If $\mathrm{Hd}(v, w)$ is odd then $\mathrm{WT}(v) \neq \mathrm{WT}(w)$.

    (b) If $\mathrm{Hd}(v, w)$ is even then $\mathrm{WT}(v) = \mathrm{WT}(w)$.

    Hint—Let $a_1, \ldots, a_n$ be the positions in which v and w differ; clearly $n = \mathrm{Hd}(v, w)$. In these positions, v and w have together a total of n ones (in addition to n zeros).

Bibliographic Note

Most of the material in this section is from the papers Kam and Ullman[48] (Theorems 1 and 2) and Chin[10] (Theorems 3 and 4). The proof of Corollary 2 can be found in Kam and Ullman,[48] the proof that $(1 + p^{-1})^{\log K}$ queries suffice on the average is contained in Chin.[10] The number of elements which must be known in order to globally compromise a database with s-queries for $s \geq 2$ is derived in Leiss.[53] For more information about graphs we refer to Harary;[37] in particular, the proof that a graph with only even-length cycles can be two colored can be found there. (This is really what we are doing in the proof of Lemma 4 when we change the query graph of DC.) The results concerning the representation of graphs as query graphs are from Leiss,[56] and the proof that every graph can be realized as a query graph can be found in Winkler;[100] the rest of this section is new material.

2.2.2. The Notion of a Tracker

Recall the database model with characteristic formulae (see the beginning of Section 2.2.1). A characteristic formula c operating on the category fields of a database determines the data values which are to be involved in the corresponding query q_c. The fact that the logical operator NOT is allowed indicates that there will be a problem not only when a formula describes records "too precisely" (e.g., if there is only one record satisfying the description) but also in the opposite situation (e.g., when there is only one record *not* satisfying the formula). Thus we will assume that the set of keys which satisfy a given formula must have at least t and at most $N - t$ elements, where N is the number of elements in the database. Clearly we have to assume

$$2 \leq t \leq N/2$$

if we want to have at least some security while at the same time providing statistical access. Contrast this situation with the one we discussed at the end of the last section. The main difference is the increased power coming from the addition of the negation operator. It will become clear soon how important this operator is for the ease and elegance with which trackers can be used to achieve compromise.

The scenario is as follows: Suppose we know—possibly from an external source—that a certain record R is uniquely identified by the formula c and suppose further that we want to determine more properties of R, for instance, whether R also possesses property a. This would be easy if we could pose the query $q_{c'}$ where

$$c' = c \text{ AND } a$$

However, $q_{c'}$ is not a permitted query since not even q_c is one. Nevertheless under certain circumstances we can in fact decide whether R satisfies a or not by using the idea of a tracker. For suppose we can write c as follows:

$$c = d \text{ AND } e$$

for certain formulae d and e such that d AND NOT e and d are valid formulae. (Here and in the following we will call a formula valid iff the corresponding query is permitted. Furthermore we will drop the "q" for query in the arguments of the functions NU and SUM.) Then the formula

$$T = d \text{ AND NOT } e$$

is called the individual tracker of R and can be used as follows to decide whether R satisfies additional properties.

Theorem 5. Let $c = d$ AND e be a formula identifying the record R and suppose $T = d$ AND NOT e is R's tracker. With three permitted queries one gets

(1) $\text{NU}(c) = \text{NU}(d) - \text{NU}(T)$ and

(2) $\text{NU}(c \text{ AND } a) = \text{NU}(T \text{ OR } d \text{ AND } a) - \text{NU}(T)$

If NU(c AND a) = 0 then R does not have property a. If NU(c AND a) = NU(c) then R has property a. Finally, if NU(c) = 1 arbitrary statistics about R can be computed using the formula

(3) $\text{SUM}(c) = \text{SUM}(d) - \text{SUM}(T)$

Proof. Since $c = d$ AND $e = d$ AND NOT T it follows that SUM(c) = SUM(d) − SUM(T) and furthermore

(4) $\text{SUM}(c \text{ AND } a) = \text{SUM}(T \text{ OR } d \text{ AND } a) - \text{SUM}(T)$

Since d AND NOT e and d are valid formulae and d AND NOT e is contained in T OR d AND a, which is itself contained in d, we can conclude that also T OR d AND a is a valid formula. This implies that all formulae on the right-hand side of equations (1), (2), (3), and (4) are valid therefore SUM(c) and SUM(c AND a) can be computed. Equations (1) and (2) are consequences of (3) and (4). □

We remark that compromise may even be possible if there is no decomposition of c which yields valid formulae d and T, for it is often possible to replace invalid formulae d and T by valid formulae d OR m and T OR m where NU(d AND m) = 0, i.e., the formula m ("mask") is only used to increase the number of records involved in order to obtain valid formulae but these new records are irrelevant as far as the actual results are concerned.

The main disadvantage of an individual tracker is that one needs a new tracker for each record. In contrast to this a general tracker works for all records in a database; furthermore no prior knowledge about any record is required.

A general tracker is any characteristic formula GT which is satisfied by no less than $2t$ and no more than N-$2t$ records. Recall that queries are permitted (and formulae valid) iff they match at least t and at most $N - t$ keys; consequently, GT is always a valid formula and the corresponding query is always permitted. Clearly GT is a general tracker iff NOT GT is a general tracker.

Theorem 6. Using a general tracker GT the value of any "not permitted" query q_c can be determined as follows. First define

$$x = \text{SUM(GT)} + \text{SUM(NOT GT)}$$

If NU(c) $< t$ then the queries on the right-hand side of the following equation are permitted:

$$\text{(5)} \quad \text{SUM}(c) = \text{SUM}(c \text{ OR GT}) + \text{SUM}(c \text{ OR NOT GT}) - x$$

If, however, NU(c) $> N - t$ then the queries on the right-hand side of this equation are permitted:

$$\text{(6)} \quad \text{SUM}(c) = 2x - \text{SUM(NOT } c \text{ OR GT}) - \text{SUM(NOT } c \text{ OR NOT GT})$$

Therefore SUM(c) can be determined in at most six queries.

Proof. As GT and NOT GT are general trackers, x can be computed. Equation (5) corresponds to the case where q_c is not permitted because less than t records satisfy c and equation (6) corresponds to the case where more than $N - t$ records are satisfied. It is clear that

$$\max(\text{NU}(c), \text{NU(GT)}) \leq \text{NU}(c \text{ OR GT}) \leq \text{NU}(c) + \text{NU(GT)}$$

If q_c matches less than t keys this inequality together with the definition of a general tracker implies

$$2t \leq \text{NU}(c \text{ OR GT}) \leq N - t$$

consequently c OR GT is valid. Similarly we show that c OR NOT GT is valid. Thus we can compute q_c in this case provided (5) holds. This, however, is easily seen from

$$((c \text{ OR GT}) \text{ AND NOT GT}) \text{ OR}$$
$$((c \text{ OR NOT GT}) \text{ AND GT}) = c$$

Let us now assume that there are more than $N - t$ keys which are matched by q_c. It follows that NOT c is satisfied by fewer than t records; consequently—using the same argument—we can compute $q(\text{NOT } c)$ using (5), namely,

$$\text{SUM(NOT } c) = \text{SUM(NOT } c \text{ OR GT)}$$
$$+ \text{ SUM(NOT } c \text{ OR NOT GT)} - x$$

Since $\text{SUM}(c) = x - \text{SUM(NOT } c)$ this yields equation (6). □

Note that no knowledge about the database is required for the use of a general tracker; even worse any formula which is satisfied by no less than $2t$ and no more the $N - 2t$ records in a database is a general tracker.

General trackers are very efficiently constructed by trial and error. Let c be a valid formula. If $2t \leq \text{NU}(c) \leq N - 2t$ then we have already a general tracker; otherwise either c or NOT(c) are satisfied by at least t and less than $2t$ elements of the database. *W*ithout *l*oss *o*f *g*enerality (WLOG) let this be c. Let d be another valid formula which is not a general tracker. Again, either $t \leq \text{NU}(d) < 2t$, or else we take NOT(d). If $\text{NU}(c \text{ OR } d) \geq 2t$ then c OR d is the desired general tracker; otherwise we continue extending our formula until we find one.

General trackers can also be systematically constructed if $t \leq N/4$. The idea is basically the same, i.e., start with a valid formula and extend it until it yields a general tracker. However, while in the method by trial and error the extension is done *ad hoc*, in the systematic construction one starts with two formulas c_1 and c_2, where c_1 is "too small" for a tracker [i.e., $t \leq \text{NU}(c_1) < 2t$] and c_2 is too big for a general tracker. Using formulas operating on the individual components of the keys one extends c_1 and restricts c_2 step by step until a general tracker is found. It follows that a general tracker can be found with a number of queries proportional to the length of the keys provided that $t \leq N/4$.

Bibliographic Note

The idea of an individual tracker is due to Schlörer.[82–84] The notion of a general tracker is introduced in Denning et al.;[24] much of the present section is based on this paper. Trackers are further studied in Denning and Schlörer,[25] where a detailed algorithm for determining a general tracker can be found.

2.3. Security in the Key-Specified Model

In this section we will discuss the security of databases where queries are key specified. We will do this for queries of several types, namely, for averages or sums, for linear sums, for arbitrary selector functions, for maxima and minima, and for medians. The section consists of two parts. In the first part (Section 2.3.1) we show that the database as defined in this model is never secure under very broad assumptions; the compromise which can be achieved ranges from (easy) global selective compromise for averages to simple compromise for arbitrary selector functions. The case of medians is interesting insofar as only partial compromise is possible, and while selective compromise can be attained it is considerably more expensive than simple compromise, which is easily achievable. We also discuss a restriction on the queries which intuitively appears to make compromise at least very difficult if not impossible; however, unfortunately our intuition turns out to be quite wrong. These results seem to indicate that it is practically impossible to have secure statistical databases. To show that this impression is nevertheless incorrect is the task we hope to accomplish in the second part (Section 2.3.2), where we give very inexpensive simple methods which guarantee the security of the underlying database. These results have the additional advantage that they are very practical, that they can be applied to any already existing statistical database without expensive modifications, and that they are provably safe.

Throughout this whole chapter we will use the following basic database model. A database is a total function

$$\mathrm{DK} : \{1, \ldots, N\} \to R$$

where N is the number of the elements in the database and R denotes the real numbers. A query is a sequence of k indices i_j in $\{1, \ldots, N\}$

where $k \geq 2$ is fixed throughout,

$$(i_1, \ldots, i_k)$$

Furthermore, $(i_1, \ldots, i_k)$ is permitted iff

$$i_j \neq i_m \qquad \text{for all } j \neq m$$

The response to such a query $(i_1, \ldots, i_k)$ of type f is

$$f(\mathrm{DK}(i_1), \ldots, \mathrm{DK}(i_k))$$

where f can be any function of k arguments.

2.3.1. Compromisable Databases

2.3.1.1. Averages and Linear Queries

2.3.1.1.1. Queries of Type Average. We first demonstrate the ease with which one can compromise databases by the use of key-specified queries of type average. We claim the following:

Theorem 7. Assume we know nothing about the elements in the database. In order to determine m arbitrary elements of the database with $N \geq \max(m, k+1)$, $\max(m, k+1)$ queries of type average and $O(m)$ arithmetic operations are sufficient (global selective compromise).

Proof Consider the following $k+1$ queries:

$$q_s = \left(\frac{1}{k}\right) * \left(\sum_{\substack{j=1 \\ j \neq s}}^{k+1} \mathrm{DK}(i_j)\right)$$

in other words, q_j is simply $(i_1, \ldots, i_{k+1})$ with i_j removed. This can be rewritten as a system of $k+1$ linear equations

$$B^* x = k^* q$$

where x is $(\mathrm{DK}(i_1), \ldots, \mathrm{DK}(i_{k+1}))$ transposed, q is $(q_1, \ldots, q_{k+1})$ transposed, and B is a $(k+1, k+1)$ matrix with the (j, s) entry equal to 1 iff $j \neq s$ and otherwise 0. It is an easy exercise to show that the matrix B

is nonsingular for all $k \geq 2$. Therefore the equation $B^*x = k^*q$ can be solved for x; hence the database can be compromised globally and selectively. This argument shows that $k+1$ queries are sufficient for $m \leq k+1$. That the claim also holds for $m > k+1$ can be seen by repeating the above argument with blocks of $k+1$ elements each (all blocks disjoint); the possibly remaining elements can be handled by using the last $k+1$ elements. Solving systems of linear equations is not a particularly attractive method; in particular we need considerably more than $O(k)$ arithmetic operations to solve a $(k+1, k+1)$ system. However, one can express the $\mathrm{DK}(i_j)$'s as follows:

$$\mathrm{DK}(i_j) = (q_1 + \cdots + q_{k+1}) - k^*q_j, \qquad j = 1, \ldots, k+1$$

By substitution one sees that this gives a solution of the equation $B^*x = k^*q$; since B is nonsingular it is the only one. Observe that $q_1 + \cdots + q_{k+1}$ need be evaluated exactly once for every set of $k+1$ elements. Hence k additions, $k+1$ subtractions, and $k+1$ multiplications by k are sufficient. An obvious extension of this argument to $m > k+1$ concludes the proof. □

Remarks. (a) It is obvious the $O(m)$ arithmetic operations are also necessary. Furthermore the number of queries for $m \geq k$ is necessary, too.

(b) It should be noted that Reiss[76] shows compromise with about 2*sqrt(k) (where sqrt denotes the square root) queries but he needs $O(k)$ arithmetic operations for determining *one* element of the database.

The method has a number of interesting aspects. For one, any error in the responses will affect only $k+1$ elements $\mathrm{DK}(i_j)$; in other words there is no indefinite propagation of errors. Round-off errors in the computation of the $\mathrm{DK}(i_j)$ also do not present any problems as the computations are numerically well behaved. However, from the point of view of data security the most alarming feature is the following:

Corollary 4. If a malicious user wants to access all the information it is irrelevant how large k is.

This implies that the intuitive feeling "The larger k, the more difficult is compromise" is not correct; computationally there is no difference in this case whether k is small or large.

Example. Let $k = 4$, and suppose we pose the following queries given together with their responses:

$$\begin{array}{rcr} (2, 3, 4, 5) & : & 2 \\ (1, 3, 4, 5) & : & -5 \\ (1, 2, 4, 5) & : & 0 \\ (1, 2, 3, 5) & : & 10 \\ (1, 2, 3, 4) & : & 3 \end{array}$$

Rewriting this gives

$$\begin{bmatrix} 0 & 1 & 1 & 1 & 1 \\ 1 & 0 & 1 & 1 & 1 \\ 1 & 1 & 0 & 1 & 1 \\ 1 & 1 & 1 & 0 & 1 \\ 1 & 1 & 1 & 1 & 0 \end{bmatrix} \cdot \begin{bmatrix} \mathrm{DK}(1) \\ \mathrm{DK}(2) \\ \mathrm{DK}(3) \\ \mathrm{DK}(4) \\ \mathrm{DK}(5) \end{bmatrix} = 4 \cdot \begin{bmatrix} 2 \\ -5 \\ 0 \\ 10 \\ 3 \end{bmatrix}$$

To determine the $\mathrm{DK}(i)$ we first compute the sum of all the responses:

$$2 - 5 + 0 + 10 + 3 = 10$$

Then we compute the $\mathrm{DK}(i)$ for $i = 1, \ldots, 5$:

$$\begin{aligned} \mathrm{DK}(1) &= 10 - 8 = 2 \\ \mathrm{DK}(2) &= 10 + 20 = 30 \\ \mathrm{DK}(3) &= 10 - 0 = 10 \\ \mathrm{DK}(4) &= 10 - 40 = -30 \\ \mathrm{DK}(5) &= 10 - 12 = -2 \end{aligned}$$

This required $k + 1 = 5$ queries and $2k + 1 = 9$ additions as well as $k + 1 = 5$ multiplications. If we want to determine $\mathrm{DK}(j)$ for $j > 5$, we have two options. Either we use the fact that we know already three elements, e.g., DK(3), DK(4), and DK(5), and that one additional query will allow us to determine $\mathrm{DK}(j)$,

$$\begin{gathered} (3, 4, 5, j) \quad : \quad q_j \\ \mathrm{DK}(j) = 4q_j - (10 - 32) = 4q_j + 22 \end{gathered}$$

Thus to determine b more elements requires b more queries and $2 + b$ additions as well as b multiplications by 4. Or we split the elements to be determined into blocks of $k + 1$ each and repeat the above for each of these blocks. While the second method requires slightly more arithmetic operations it has the advantage that an error in the computation of any of the elements as well as of the queries (e.g., rounding error) will affect at most $k + 1$ elements. In contrast, an error in the computation of the elements DK(3), DK(4), or DK(5) in the first method will affect all of the subsequent computations.

We close with a remark on queries of type geometric mean, i.e., queries where the response is given by

$$(\mathrm{DK}(i_1)^*\mathrm{DK}(i_2) \ldots {}^*\mathrm{DK}(i_k))^{1/k}$$

If it is known that all the values of the database are positive then the present model with averages is isomorphic to this model of geometric means by virtue of the functions exponentiation and logarithm. If the values are not all positive additional ambiguities are introduced which may not be resolvable at all.

Example. Assume that queries are of type geometric mean, and let $k = 2$. Consider the following queries and their responses:

$$\begin{array}{ccc} (2,3) & : & 2 \\ (1,3) & : & 1 \\ (1,2) & : & 12 \end{array}$$

Applying logarithms yields the following:

$$\begin{bmatrix} 0 & 1 & 1 \\ 1 & 0 & 1 \\ 1 & 1 & 0 \end{bmatrix} \cdot \begin{bmatrix} \log[\mathrm{DK}(1)] \\ \log[\mathrm{DK}(2)] \\ \log[\mathrm{DK}(3)] \end{bmatrix} = 2 \cdot \begin{bmatrix} \log 2 \\ \log 1 \\ \log 12 \end{bmatrix}$$

Thus we get

$$\begin{aligned} \log[\mathrm{DK}(1)] &= (\log 2 + \log 1 + \log 12) - 2 \cdot \log 2 \\ \log[\mathrm{DK}(2)] &= (\log 2 + \log 1 + \log 12) - 2 \cdot \log 1 \\ \log[\mathrm{DK}(3)] &= (\log 2 + \log 1 + \log 12) - 2 \cdot \log 12 \end{aligned}$$

and from this we get by applying exponentiation

$$\begin{aligned} \mathrm{DK}(1) &= 24/4 = 6 \\ \mathrm{DK}(2) &= 24/1 = 24 \\ \mathrm{DK}(3) &= 24/144 = 1/6 \end{aligned}$$

However, it should be noted that applying logarithms in this way is possible only if it is known that all the numbers are positive. If this is not the case the solution of these systems of equations may not be unique, consequently it is not possible to determine the *correct* values. For example, the same responses would be obtained with the following values:

$$\mathrm{DK}(1) = -6$$
$$\mathrm{DK}(2) = -24$$
$$\mathrm{DK}(3) = -1/6$$

Even worse, if any of the responses is zero this method cannot be applied at all. However, it may be possible to compromise nevertheless as for a response to be zero at least one of the $\mathrm{DK}(i)$ involved in the query must be zero. By posing additional queries one may be able to determine which of the elements has the value zero.

2.3.1.1.2. Queries of Type Linear Sum. A linear sum with weights $a_1, \ldots, a_k$ is a function in the variables $x_1, \ldots, x_k$ of the form

$$a_1 x_1 + \cdots + a_k x_k$$

thus we are dealing with queries $(i_1, \ldots, i_k)$ whose response is given by

$$a_1 \cdot \mathrm{DK}(i_1) + \cdots + a_k \cdot \mathrm{DK}(i_k)$$

The weights $a_1, \ldots, a_k$ are assumed to be arbitrary and fixed throughout. Clearly this is a major generalization of queries of type average for which the weights are all $1/k$.

In order to prepare the main result about queries of type linear sum we first show the following:

Proposition 3. Knowing either one weight and one data element or two data elements is sufficient to compromise globally and selectively.

Proof. Consider the queries

$$(i_1, i_2, \ldots, i_{j-1}, s, i_{j+1}, \ldots, i_k)$$

and

$$(i_1, i_2, \ldots, i_{j-1}, t, i_{j+1}, \ldots, i_k)$$

with responses q_1 and q_2, respectively, where $\mathrm{DK}(s)$ is the data element we know. Subtracting yields

$$q_1 - q_2 = a_j{}^*(\mathrm{DK}(s) - \mathrm{DK}(t))$$

and from this we can either determine a_j if DK(t) is the second known data element, or we can determine DK(t) if a_j is the known weight. [We assume that $q_1 = q_2$ and DK(s) = DK(t) do not hold at the same time. If this assumption is not met, one has to choose different queries.] Now it is obvious how the whole database can be compromised; we simply repeat the above argument for other data elements in place of DK(t). □

Note that this method allows us to determine any of the weights.

Example. Let $k = 3$, and assume that we know one weight and one element of the database, i.e.,

$$a_1 = 1/4 \qquad \text{and} \qquad \mathrm{DK}(1) = 12$$

In order to determine the remaining weights we first determine one additional data element from the two queries

$$(1, 3, 4) \quad : \quad 5.0$$
$$(2, 3, 4) \quad : \quad 9.0$$

it follows that DK(2) = 28.0 as

$$9.0 - 5.0 = \mathrm{DK}(2)/4 - 12.0/4$$

Now we can determine a_2 from the following queries:

$$(3, 2, 4) \quad : \quad 6.4$$
$$(3, 1, 4) \quad : \quad 0.0$$

as $6.4 - 0.0 = a_2 \cdot (28.0 - 12.0)$ we can conclude that

$$a_2 = 2/5$$

Finally,

$$(3, 4, 2) \quad : \quad 2.2$$

and

$$(3, 4, 1) \quad : \quad -3.4$$

allow us to conclude that

$$a_3 = 7/20$$

Consequently, $a_1 = 1/4$, $a_2 = 2/5$, and $a_3 = 7/20$.

Proposition 3 assumes very little knowledge on the part of the user about the database. While it is rather unrealistic to assume that a user does not know any of the elements it may be not so unreasonable to assume that none of the weights is known. However, even in the case where only one data element is known but no weights, compromise is possible as is stated in the following:

Theorem 8. Suppose that one data element is known. It may be possible to determine k additional data elements with no more than $k(k+1)$ queries. Global compromise is then possible with altogether $N + k^2 - 1$ queries.

Proof. Assume we know the data element DK(1) and we wish to determine the elements DK(2), ..., DK($k+1$). Define $Z = (1, \ldots, k+1)$ and let Z_j be Z with j deleted. Let $\mathbf{r}$ be the cyclic permutation on k objects given by

$$\mathbf{r}: \begin{matrix} 1 & 2 & 3 & \ldots & k-1 & k \\ 2 & 3 & 4 & \ldots & k & 1 \end{matrix}$$

Then define $Z(j, 0)$ to be Z_j and $Z(j, t+1) = \mathbf{r}(Z(j, t))$ for $t = 0, \ldots, k-2$, and let $q(j, t)$ be the response to the query $Z(j, t)$. As $0 \leq t \leq k-1$ and $1 \leq j \leq k+1$, exactly $k(k+1)$ queries are used. Let $S = \mathrm{DK}(1) + \cdots + \mathrm{DK}(k+1)$; hence the sum of all the elements involved in $q(j, 0)$ is $S - \mathrm{DK}(j)$, and furthermore

$$\begin{aligned} &q(j, 0) + \cdots + q(j, k-1) \\ &\quad = a_1 \cdot [S - \mathrm{DK}(j)] + \cdots + a_k \cdot [S - \mathrm{DK}(j)] = A \cdot [S - \mathrm{DK}(j)], \end{aligned}$$

where $A = a_1 + \cdots + a_k$; let us denote this value by p_j:

$$p_j = A \cdot [S - \mathrm{DK}(j)] \qquad \text{for } j = 1, \ldots, k+1$$

Assume now that $S - \mathrm{DK}(j) \neq 0$ for all $j = 1, \ldots, k$. Then we obtain

$$A = p_j/[S - \mathrm{DK}(j)] = p_{j+1}/[S - \mathrm{DK}(j+1)] \qquad \text{for } j = 1, \ldots, k$$

and from this we can eliminate A. This yields a system of k linear equations in the k unknowns DK(2), . . . , DK($k+1$), namely, $S^*(p_j - p_{j+1}) + p_{j+1}{}^*\text{DK}(j) - p_j{}^*\text{DK}(j+1) = 0$, $j = 1, \ldots, k+1$. Let us furthermore assume that these equations are linearly independent. Then we can determine the variables DK(2), . . . , DK($k+1$) by solving this system of linear equations. Now by Proposition 3 we can determine the weights, and from there again by Proposition 3 we can determine all other data elements with one additional query per new data element. □

It should be pointed out that the security of the database is not guaranteed if the above assumption about the linear independence of the equations is violated as there might be some other choice of DK(i_2), . . . , DK(i_{k+1}) in place of the DK(2), . . . , DK($k+1$) which does result in a system of linearly independent equations. The same holds for the assumption that $S - \text{DK}(j) \neq 0$ for all $j = 1, \ldots, k$.

Example. Let $k = 3$, and assume that

$$\text{DK}(1) = 3$$

is known. We want to determine DK(2), DK(3), and DK(4). No weights are known. The queries $Z(j, t)$ and their responses $q(j, t)$ are as follows:

$$Z(1,0) = (2,3,4) \quad : \quad 10.0$$
$$Z(2,0) = (1,3,4) \quad : \quad 9.5$$
$$Z(3,0) = (1,2,4) \quad : \quad 8.5$$
$$Z(4,0) = (1,2,3) \quad : \quad 7.0$$

$$Z(1,1) = (3,4,2) \quad : \quad 8.5$$
$$Z(2,1) = (3,4,1) \quad : \quad 7.0$$
$$Z(3,1) = (2,4,1) \quad : \quad 6.5$$
$$Z(4,1) = (2,3,1) \quad : \quad 5.5$$

$$Z(1,2) = (4,2,3) \quad : \quad 8.5$$
$$Z(2,2) = (4,1,3) \quad : \quad 7.5$$
$$Z(3,2) = (4,1,2) \quad : \quad 6.0$$
$$Z(4,2) = (3,1,2) \quad : \quad 5.5$$

Consequently, by adding up $q(j, 0) + \cdots + q(j, k-1)$ we get p_j for all $j = 1, \ldots, k+1$:

$$p_1 = 27.0$$
$$p_2 = 24.0$$
$$p_3 = 21.0$$
$$p_4 = 18.0$$

From this we get the following system of linear equations:

$$\begin{bmatrix} p_1 & -p_2 & p_1 - p_2 & p_1 - p_2 \\ p_2 - p_3 & p_2 & -p_3 & p_2 - p_3 \\ p_3 - p_4 & p_3 - p_4 & p_3 & -p_4 \end{bmatrix} \cdot \begin{bmatrix} \mathrm{DK}(1) \\ \mathrm{DK}(2) \\ \mathrm{DK}(3) \\ \mathrm{DK}(4) \end{bmatrix} = \begin{bmatrix} 0 \\ 0 \\ 0 \end{bmatrix}$$

or

$$\begin{bmatrix} 27 & -24 & 3 & 3 \\ 3 & 24 & -21 & 3 \\ 3 & 3 & 21 & -18 \end{bmatrix} \cdot \begin{bmatrix} \mathrm{DK}(1) \\ \mathrm{DK}(2) \\ \mathrm{DK}(3) \\ \mathrm{DK}(4) \end{bmatrix} = \begin{bmatrix} 0 \\ 0 \\ 0 \end{bmatrix}$$

Recalling that we know $\mathrm{DK}(1) = 3$ we can rewrite this as

$$\begin{bmatrix} -24 & 3 & 3 \\ 24 & -21 & 3 \\ 3 & 21 & -18 \end{bmatrix} \cdot \begin{bmatrix} \mathrm{DK}(2) \\ \mathrm{DK}(3) \\ \mathrm{DK}(4) \end{bmatrix} = \begin{bmatrix} -81 \\ -9 \\ -9 \end{bmatrix}$$

If follows immediately that this system of equations is nonsingular and the solution is as follows:

$$\mathrm{DK}(2) = 6, \qquad \mathrm{DK}(3) = 9, \qquad \mathrm{DK}(4) = 12$$

It should be noted that only now we can verify the second assumption made in the proof, namely, that $S - \mathrm{DK}(j) \neq 0$ for all $j = 1, \ldots, k+1$. In our example we have $S = 30$ and hence $S - 3 \neq 0$, $S - 6 \neq 0$, $S - 9 \neq 0$, and $S - 12 \neq 0$. Hence the assumption is valid in our case. Note that it is conceivable that one manages to "solve" the resulting system of equations but then finds out that not all $S - \mathrm{DK}(j)$ are nonzero. In this case one can substitute the computed values for the $\mathrm{DK}(j)$ into the queries in order to determine whether they actually satisfy the queries. Even if this is the case there is no absolute guarantee that these are the database elements; however, it should be easy to ascertain this with a few additional queries. Note that by Proposition 3 it is now easy to find the weights, in fact without additional queries. In our example one finds

$$a_1 = 1/6, \qquad a_2 = 1/3, \qquad a_3 = 1/2$$

For the sake of completeness let us now consider the case where we know one of the weights but none of the data elements.

Theorem 9. Suppose that one weight is known but no data element. It may be possible to determine $k+1$ data items with no more than $k(k+1)$ queries. Global compromise is then possible with $N+k^2-1$ queries.

Proof. The key idea is to use the proof of Theorem 8 pretending that we know one data element. Then we can use the one known weight to determine this data element and the result follows.

Assume DK(1) to be known and apply Theorem 8. This yields

$$\text{DK}(j) = F_j(\text{DK}(1)) \qquad \text{for } j = 2, \ldots, k+1$$

where F_j is a function of the variable DK(1) only. (In fact it is not too difficult to see that F_j is a linear function in this variable.) Using this one can also express the weights in terms of functions of DK(1) only, namely,

$$a_j = G_j(\text{DK}(1)) \qquad \text{for } j = 1, \ldots, k$$

Since one of the a_j is known it follows that DK(1) can be determined. Then the result follows by Theorem 8. □

Again Theorem 9 does not guarantee compromise but it does show that under certain circumstances it is possible.

One can prove that without any knowledge about the database (neither a data element nor a weight) compromise cannot be achieved. However, as this assumption is virtually never met in practice we will refer to the original paper for this proof. For all practical purposes this kind of database must be considered compromisable.

Example. Suppose we know the value of the weight a_1, say, $a_1 = 1/6$, but we do not know any data element. We first start out as we did in the last example until we get to the system of equations

$$\begin{bmatrix} 27 & -24 & 3 & 3 \\ 3 & 24 & -21 & 3 \\ 3 & 3 & 21 & -18 \end{bmatrix} \cdot \begin{bmatrix} \text{DK}(1) \\ \text{DK}(2) \\ \text{DK}(3) \\ \text{DK}(4) \end{bmatrix} = \begin{bmatrix} 0 \\ 0 \\ 0 \end{bmatrix}$$

Now we *assume* that DK(1) is known; this yields

$$\begin{bmatrix} -24 & 3 & 3 \\ 24 & -21 & 3 \\ 3 & 21 & -18 \end{bmatrix} \cdot \begin{bmatrix} \mathrm{DK}(2) \\ \mathrm{DK}(3) \\ \mathrm{DK}(4) \end{bmatrix} = \begin{bmatrix} -27\mathrm{DK}(1) \\ -3\mathrm{DK}(1) \\ -3\mathrm{DK}(1) \end{bmatrix}$$

By the usual matrix manipulation this can be transformed into

$$\begin{bmatrix} -2 & 1 & 0 \\ 8 & -7 & 1 \\ 1 & 0 & 0 \end{bmatrix} \cdot \begin{bmatrix} \mathrm{DK}(2) \\ \mathrm{DK}(3) \\ \mathrm{DK}(4) \end{bmatrix} = \begin{bmatrix} -\mathrm{DK}(1) \\ -\mathrm{DK}(1) \\ 2\mathrm{DK}(1) \end{bmatrix}$$

and this yields

$$\begin{aligned} \mathrm{DK}(2) &= 2\mathrm{DK}(1) \\ \mathrm{DK}(3) &= 3\mathrm{DK}(1) \\ \mathrm{DK}(4) &= 4\mathrm{DK}(1) \end{aligned}$$

Using, for example, the queries $Z(1, 0)$ and $Z(2, 0)$ we determine the weight a_1 by the method of Proposition 3:

$$0.5 = 10.0 - 9.5 = a_1[\mathrm{DK}(2) - \mathrm{DK}(1)]$$

and solving this equation for a_1 gives

$$a_1 = 1/[2 \cdot \mathrm{DK}(1)]$$

However, since a_1 is known, namely, $a_1 = 1/6$, this allows us to determine the value of DK(1) to be 3. Now everything else follows by Theorem 8. Note that the verification of the assumptions must again be performed. In the present case, however, there is no problem, thus the database can be globally compromised.

Exercises

1. Determine the maximal error occurring when compromising a database according to the method given in the proof of Theorem 7. More specifically, assume floating point numbers with d significant digits and express the error in terms of d and k.

2. Compare the maximal errors of the suggested methods for determining elements of a database (let m be a multiple of $k + 1$), namely,

continuous computation of $DK(j)$ for $y = k + 2, \ldots, m$ using the already computed values for $DK(1), \ldots, DK(k + 1)$ (which of course may be afflicted with errors), and blocking.

3. Prove that the matrix $B = (b_{ij})$ is nonsingular for all $k \geq 2$, where $b_{ij} = 1$ if $i \neq j$ and $b_{ij} = 0$ if $i = j$.

4. Show that the following method achieves compromise with about 2sqrt(k) queries of type average. Let $k = h^2 + 1$, and let for all i in $\{1, \ldots, 2h + 1\}$

$$y_i = DK(h(i - 1) + 1) + \ldots + DK(h(i - 1) + h)$$

Then the queries are

$$DK(2h^2 + h + 1) + (y_1 + \cdots + y_h)$$

$$(y_1 + \cdots + y_h) - y_i + y_{k+i+1} + DK(h^2 + i) \qquad \text{for } 1 \leq i \leq h$$

$$(y_{h+1} + \cdots + y_{2h}) - y_i + DK(2h^2 + h + 1) \qquad \text{for } h + 1 \leq i \leq 2h + 1$$

Now multiply the first query by $h(h - 1)$, multiply the second group of queries by $-h$, and add the three groups together.

5. Show that queries of type geometric mean do not permit compromise provided k is even and the entries of the database are positive and negative (i.e., $\neq 0$). Hint—Assume that $x_1, \ldots, x_m$ with $x_i > 0$ is a solution obtained under the assumption that all entries of the database are positive. Then $-x_1, \ldots, -x_m$ will give the same responses.

6. Assume you know that all entries in the database are non-negative (i.e., positive and zero). Assume that there are z nonzero elements in the database. Let the queries be of type geometric mean of k arguments.

   (a) Prove that for $z < k$, compromise is impossible.

   (b) Show that for $z \geq k + 1$, all (zero and positive!) elements of the database can be determined.

   (c) Give an algorithm to achieve the global compromise of question (b).

   (d) What is the best upper bound for the worst case of determining *one* entry of this database (assuming $z > k$)?

(e) Show that compromise can be achieved in the best case with exactly two queries (independent of k). Hint to (e)—Consider two queries differing in just one element, one responding with 0, the other with a reply $\neq 0$.

7. Can you compromise the database assuming you know that

$$\begin{aligned} \mathrm{cbrt}(\mathrm{DK}(2) \cdot \mathrm{DK}(3) \cdot \mathrm{DK}(4)) &= -4 \\ \mathrm{cbrt}(\mathrm{DK}(1) \cdot \mathrm{DK}(3) \cdot \mathrm{DK}(4)) &= 4 \\ \mathrm{cbrt}(\mathrm{DK}(1) \cdot \mathrm{DK}(2) \cdot \mathrm{DK}(4)) &= 2 \\ \mathrm{cbrt}(\mathrm{DK}(1) \cdot \mathrm{DK}(2) \cdot \mathrm{DK}(3)) &= -2 \end{aligned}$$

where cbrt denotes the cubic root?

8. Expand the method you used in Exercise 7 to obtain a general method for odd k for database entries which may be positive and negative (but not zero).

9. Suppose you are given the following queries of type weighted average:

2 3 4 5	7.25	1 3 4 5	5.25	1 2 4 5	4.25
1 2 3 5	3.5	1 2 3 4	3.25	3 4 5 2	10.25
3 4 5 1	10.0	2 4 5 1	8.0	2 3 5 1	7.0
2 3 4 1	6.25	4 5 2 3	11.25	4 5 1 3	10.5
4 5 1 2	10.25	3 5 1 2	8.25	3 4 1 2	7.25
5 2 3 4	11.25	5 1 3 4	10.25	5 1 2 4	9.5
5 1 2 3	9.25	4 1 2 3	7.25		

Can you compromise this database if you know (a) that DK(1) = 0? (b) that the first weight $a_1 = 1/2$?

10. Same question as in Exercise 9 but the responses to all the queries are 1 and (a) you know that DK(1) = 1 (b) you know that weight $a_1 = 1/9$.

11. What is the minimum number of queries necessary in the worst case to detect that *all* of $k + 1$ database entries are pairwise equal? What are these queries? Hint—What can you infer about $q(j, t)$ if all the elements are equal [cf. 10(a)]? Then consider $Z(j, 0), j = 1, \ldots, k + 1$.

Bibliographic Note

Theorem 7 has been independently obtained by several authors, among them Davida *et al.*[17] and by Schwartz;[86] the version given here follows Leiss.[51] Geometric means are mentioned in Leiss.[51] Linear queries are considered in Schwartz *et al.*[87]; the result concerning compromise if one data element is known (Theorem 8) can be found there as well as the proof that compromise is not possible if nothing is known about the database.

2.3.1.2. Restricted Overlap for Averages

In the last section we saw that key-specified queries of type average must not be allowed if one wants to protect the security of the data stored. However, it should be clear that precisely these kinds of queries are used very frequently in actual statistical databases. This makes it mandatory to search for modifications of, or restrictions on, these queries such that the resulting queries satisfy two requirements:

(a) Their response should be precisely the average of the data elements indicated in the query.

(b) It should be impossible to compromise the database with these queries.

Restricting the overlap between queries is a scheme which has been suggested as a solution to this problem. In this section we will define this model and discuss its feasibility.

Suppose we are given a sequence of m queries

$$(i_{j,1}, \ldots, i_{j,k}) \qquad \text{for } j = 1, \ldots, m$$

with response $q_j = 1/k(\mathrm{DK}(i_{j,1}) + \cdots + \mathrm{DK}(i_{j,k}))$. We say this sequence has overlap r if any two queries have no more than r indices in common, i.e., for all s, t in $\{1, \ldots, m\}$, $s \neq t$, we have

$$\mathrm{card}(\{i_{s,1}, \ldots, i_{s,k}\} \cap \{i_{t,1}, \ldots, i_{t,k}\}) \leq r$$

While in this section we consider only queries of type average it should be obvious that the definition of restricted overlap applies to queries of any type as the restriction can be defined exclusively in terms of the indices involved only. We can restrict r to be less than k and at least 0:

$$0 \leq r \leq k - 1$$

Clearly $r > k$ is impossible as is $r < 0$; $r = 0$ means that no overlap is permitted, and $r = k$ would indicate the repetition of a query, which of course in the present setting has no effect on the security of the database considered. In fact it will always be permitted to repeat a query even if we restrict the overlap r to be less than k. The methods in Section 2.3.1.1 for achieving compromise required overlap $r = k - 1$, the largest possible which contributes new information. This suggests that it may be at least much more difficult or even impossible to compromise with small overlap. We will show in the following that this is indeed the case; briefly we can say the smaller r the more difficult it is to compromise, from $r = k - 1$, where we know already that it is very easy, to $r = 0$, where it is impossible to compromise. However, there are very serious objections to this method which render it unfeasible for all practical purposes. These problems will be pointed out later. First, however, we will discuss the method in more detail.

In the following we will always assume that no data elements are known:

$$D_0 = \emptyset$$

Clearly if some elements are in fact known this will only decrease the security of the database. Since we will show that the method is not very useful anyway this assumption does not change the implications of the results.

First let us get a feeling for the effect of the restriction on the overlap of the queries in general.

Theorem 10. Let $D_0 = \emptyset$. Selective compromise is possible using queries of type average with overlap at most 1, provided the database contains at least $k(k - 1) + 1$ elements. Exactly $2k - 1$ queries are necessary and sufficient for this compromise.

Proof. There are two parts we have to show, namely, that we can compromise with $2k - 1$ queries and that we have to use at least that many queries. First we prove sufficiency. Consider the following sequence of queries $q_1, \ldots, q_{2k-1}$; as before we write for simplicity j instead of i_j:

$$q_s : (k(s-1)+1, \ldots, k(s-1)+k) \qquad \text{for } s = 1, \ldots, k-1$$

$$q_{s+k-1} : (k(1-1)+s, k(2-1)+s, \ldots, k(k-1-1)+s, k(k-1)+1) \qquad \text{for } s = 1, \ldots, k$$

This can be pictured as follows:

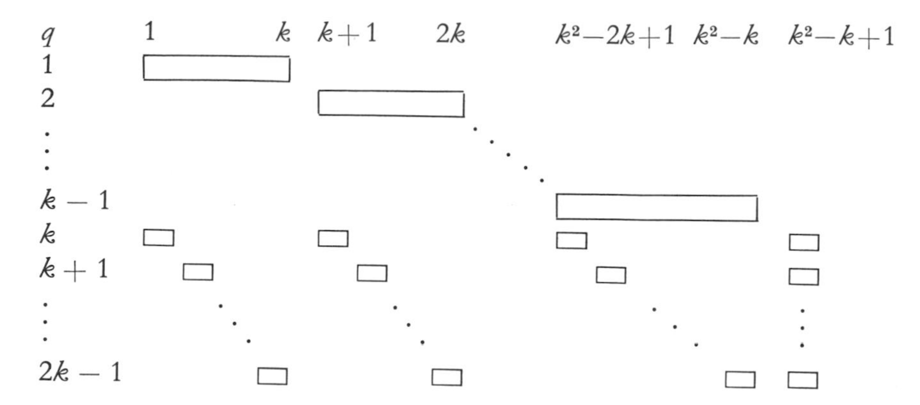

By inspection one verifies that indeed these queries have overlap at most 1. It follows easily that adding up the first $k-1$ queries and subtracting from this all remaining k queries cancels out all indices but the last one, $k(k-1)+1$. Hence we have

$$-\mathrm{DK}(k(k-1)+1) = (q_1 + \cdots + q_{k-1} - q_k - \cdots - q_{k-1})/k$$

Therefore it is possible to compromise with $2k-1$ queries of overlap at most 1.

Now we prove necessity. Suppose we have a shortest compromising sequence of say t queries; without loss of generality we assume that we can determine $\mathrm{DK}(i_1)$. We can write this sequence as follows:

$$q_j : (i_{j,1}, \ldots, i_{j,k}) \qquad \text{for } j = 1, \ldots, t$$

where

$$1 \leq i_{j,1} < \cdots < i_{j,k} \leq N \qquad \text{for } j \text{ in } \{1, \ldots, t\}$$

and

$$\mathrm{card}(\{i_{j,1}, \ldots, i_{j,k}\} \cap \{i_{m,1}, \ldots, i_{m,k}\}) \leq 1 \qquad \text{for } j \neq m$$

That we can determine that $\mathrm{DK}(i_1)$ is represented by

$$\mathrm{DK}(i_1) = c_1 q_1 + \cdots + c_t q_t$$

where

$$c_j \neq 0 \qquad \text{for all } j \text{ in } \{1, \ldots, t\}$$

Rewriting this sum we obtain

$$\mathrm{DK}(i_1) = \sum_{j=1}^{t} c_j q_j = \sum_{j=1}^{t} c_j \cdot \left[\sum_{s=1}^{k} \mathrm{DK}(i_{j,s}) \right] = \sum_{m=1}^{N} \left(\sum_{j=1}^{t} c_j d_{j,m} \right) \cdot \mathrm{DK}(m)$$

where $d_{j,m} = 1$ if $\mathrm{DK}(m)$ is used in query q_j and $d_{j,m} = 0$ otherwise, for j in $\{1, \ldots, t\}$ and m in $\{1, \ldots, N\}$. It follows that for all $m \neq i_1$ the term $c_1 d_{1,m} + \cdots + c_t d_{t,m}$ must be 0 and for $m = i_1$ it must be 1. Note that all the elements of the database can be treated as indeterminates (which therefore are all linearly independent) since this method works independently of the values of the $\mathrm{DK}(m)$; in other words if we change the values of the $\mathrm{DK}(m)$ the terms $c_1 d_{1,m} + \cdots + c_t d_{t,m}$ are not affected. In order for this sum to be 0 there are two possibilities: either all $d_{j,m}$'s involved are 0, or there are $d_{s,m}$ and $d_{n,m}$ which are equal to 1, in which case c_j and c_n must have opposite sign (recall that $c_j \neq 0$). This implies that any $\mathrm{DK}(m)$ which appears in some query must in fact appear in at least two queries, for $m \neq i_1$, and the coefficients c_j of these queries must have opposite signs.

Let us now determine a lower bound on t. Obviously at least one query, say, q_1, is needed, i.e., $t \geq 1$, and let WLOG $c_1 > 0$. Furthermore we may assume that $\mathrm{DK}(i_1)$ appears in q_1. Since the first query assigns to k of the $\mathrm{DK}(m)$'s a positive c_1, at least one query, q_2, with a negative c_2 is necessary. Let T_1 be 1 if $\mathrm{DK}(i_1)$ appears also in the second query and $T_1 = 0$ otherwise. Let T_2 be the number of $\mathrm{DK}(m)$'s for $m \neq i_1$ which appear in both queries. Then because of the first query we need at least another $k - T_2 - 1$ additional queries with negative c_j, and because of the second query we need at least another $k - T_1 - T_2$ additional queries with a positive c_j. However,

$$T_2 + T_1 \leq 1 = r$$

and therefore

$$t \geq 2 + k - T_1 + k - T_2 - 1 - T_2 \geq 2k - 1$$

This concludes the proof of necessity; consequently the theorem is proven. □

Example. Let $k = 4$; therefore we must have at least 13 database elements. The queries resulting from the method above are given in the following array:

Response	q	1	2	3	4	5	6	7	8	9	10	11	12	13	Factor
5	1	1	1	1	1										--1
−1	2					1	1	1	1						−1
5	3									1	1	1	1		−1
7	4	1				1				1				1	+1
0	5		1				1				1			1	+1
−9	6			1				1				1		1	+1
1	7				1				1				1	1	+1

It follows that

$$4\mathrm{DK}(13) = -(5 - 1 + 5) + (7 + 0 - 9 + 1) = -8$$

or

$$\mathrm{DK}(13) = -2$$

Theorem 10 gives a precise description of the situation if we want to determine just one element. If we wanted to know the values of more elements we could repeat this method. However, this is quite inefficient. In what follows we will outline a method whereby one can determine $k(k - 1) + 1$ elements using $k(k - 1) + 1$ queries of type average with overlap restricted to be at most 1 provided the database contains at least $k(k - 1) + 1$ elements,

$$N \geq k(k - 1) + 1$$

In the derivation of this result we will make use of some theorems concerning symmetric block designs. However, there is no need for us to give an outline of the theory of symmetric block designs; the interested reader is referred to M. Hall's book[36] for more information.

Let us consider sequences Q_t of t queries of type average where any two queries have precisely(!) r indices in common. Now suppose that we can find a (t, t) matrix $E_{k,r}$ where all entries are 0 or 1, any row has exactly k ones, and any two rows have exactly r ones in the same column. Clearly such a matrix $E_{k,r}$ gives rise immediately to a sequence of t queries with overlap r. Such matrices have a number of interesting properties, among them the following:

$$E_{k,r}{}^{*}F_{k,r} = (k - r)^{*}I + r^{*}J \tag{1}$$

where F is the transpose of E, I is the identity matrix, and J is the matrix consisting of ones only. The equation can be proven by considering the inner product of a row of $E_{k,r}$ with itself:

$$\det(E_{k,r} * F_{k,r}) = (k - r)^{t-1} \cdot (tr - r + k) \quad (2)$$

For a proof see Hall (Reference 36, p. 103). Using this equation and the fact that

$$t > k > r$$

one deduces that $E_{k,r}$ is nonsingular for any choice of k and r:

$$E_{k,r} * J = k * J \quad (3)$$

We now use a theorem by Ryser (Reference 36, p. 104, Theorem 10.2.3) which states that (1), (2), and (3) imply

$$t = k(k - 1)/r + 1$$

Thus in particular matrices $E_{k,r}$ exist only if r divides $k(k - 1)$. Moreover using the Bruck–Ryser–Chowla theorem (Reference 36, p. 107, Theorem 10.3.1) and its partial converse, which applies in our case (Reference 36, p. 282, Theorem 16.4.1, and Sections 10.3 and 10.4) we get the following:

Lemma 5. For given k and r ($k > r$), a (t, t) matrix $E_{k,r}$ exists iff (1) $t = k(k - 1)/r + 1$, and (2) t is even and $k - r$ is a perfect square or t is odd and the following equation has a solution in integers x, y, z different from (0, 0, 0):

$$z = (k - r) \cdot x^2 + (-1)^{(t-1)/2} \cdot r \cdot y$$

It can be shown (Reference 36, p. 111) that the equation in (2) always has a solution for $r = 1$. Thus we can state the following:

Corollary 5. For any $k \geq 2$ there exists a (t, t) matrix $E_{k,1}$ with $t = k(k - 1) + 1$.

Now we can relate these results to the security of statistical databases, namely, we can prove the following:

Theorem 11. Let $D_0 = 0$. Suppose we allow nonzero overlap in queries of type average. If the database has at least $k(k - 1) + 1$ elements then it can be globally and selectively compromised. Furthermore

$k(k-1)+1$ queries suffice to determine that many elements, or N queries to determine all elements.

Proof. By Corollary 5 there exist matrices $E_{k,1}$ for all $k \geq 2$. Thus let Q_t be the sequence of queries corresponding to $E_{k,1}$ with $t = k(k-1)+1$. As $E_{k,1}$ is nonsingular we can solve the system of linear equations

$$E_{k,1} {}^* x = k^* q$$

where x is $(\mathrm{DK}(i_1), \ldots, \mathrm{DK}(i_t))$ transposed, q is $(q_1, \ldots, q_t)$ transposed, and q_j is the response to the jth query. Thus any of the $\mathrm{DK}(i_j)$ can be computed; the database can be globally and selectively compromised. Finally, once we know at least $k-1$ elements it is completely trivial to determine other values using one query per unknown element. □

We point out that Hall (Reference 36, Section 10.3) gives a method for constructing a matrix $E_{k,1}$ for given k. However, as we are not so much interested in methods for compromising a database but rather in schemes which guarantee the security of a database, the proof that compromise is possible within this setting will suffice for us here.

Example. Suppose $k=3$; furthermore overlap between any two queries must not be greater than 1. We first construct a matrix $E_{3,1}$ of dimension $k(k-1)/1+1=7$:

$$\begin{bmatrix} 1 & 1 & 1 & 0 & 0 & 0 & 0 \\ 1 & 0 & 0 & 1 & 1 & 0 & 0 \\ 0 & 1 & 0 & 1 & 0 & 1 & 0 \\ 0 & 0 & 1 & 0 & 1 & 1 & 0 \\ 1 & 0 & 0 & 0 & 0 & 1 & 1 \\ 0 & 1 & 0 & 0 & 1 & 0 & 1 \\ 0 & 0 & 1 & 1 & 0 & 0 & 1 \end{bmatrix}$$

Then we pose the following seven queries:

1, 2, 3	:	3
1, 4, 5	:	2
2, 4, 6	:	1
3, 5, 6	:	0
1, 6, 7	:	−1
2, 5, 7	:	−2
3, 4, 7	:	−3

Solving the corresponding system of linear equations yields the following results:

$$\begin{aligned} DK(1) &= 6 \\ DK(2) &= 3 \\ DK(3) &= 0 \\ DK(4) &= 0 \\ DK(5) &= 0 \\ DK(6) &= 0 \\ DK(7) &= -9 \end{aligned}$$

Theorem 11 alone would suggest that restricted overlap is not a very feasible method. But let us nevertheless assume someone wanted to implement it. The following problems would arise. In order to be able to test whether the overlap condition is satisfied all queries of a user must be stored; note that there can be almost arbitrarily many as disjoint queries do not affect anything. In fact these queries must be stored at least as long as the information in the database remains substantially unchanged. In the case of a large database with many users a prohibitively large amount of (practically always useless) information must be stored for every user. Moreover, any query ever issued must be compared against all the queries previously issued in order to test for the overlap condition. But even if we were willing to do all this we still could not guarantee the security of our database: Two or more malicious users could (perfectly legally) combine their results thereby circumventing our elaborate mechanisms. Note that these arguments also apply if one restricted the overlap to be zero. (Clearly if all queries issued by all users have overlap zero the security of the database is in fact guaranteed!)

In summary using the method of restricted overlap in order to protect the security of a statistical database is totally infeasible from a practical point of view because the conditions under which it would work are unenforceable.

Exercises

1. Show that $E_{k,r}*F_{k,r} = (k-r)*I + r*J$ with F being the transpose of E, I the identity matrix, and J the matrix consisting of ones only.

2. Show that $E_{k,r}*J = k*J$.

3. Construct matrices $E_{k,r}$ for (a) $r = k - 1$ and arbitrary k, (b) $k = 5$ and $r = 2$, (c) $k = 6$ and $r = 2$, (d) $k = 7$ and $r = 3$, (e) $k = 8$ and $r = 4$, (f) $k = 10$ and $r = 6$, (g) $k = 9$ and $r = 5$.

4. Suppose $k = 4$ and the overlap $r = 2$. Furthermore assume the following queries are known:

$$\begin{aligned} 1, 2, 3, 4 &\ :\ 18 \\ 1, 2, 5, 6 &\ :\ 14 \\ 3, 4, 5, 6 &\ :\ 10 \\ 1, 4, 5, 7 &\ :\ 11 \\ 2, 3, 5, 7 &\ :\ 11 \\ 1, 3, 6, 7 &\ :\ 11 \end{aligned}$$

Can you compromise the database with these queries? If not, can you add queries in order to compromise? How many do you need? Let the response to additional queries be the sum of the indices *not* present in the query, i.e., if the query is (a, b, c, d) with $1 \leq a, b, c, d \leq 7$ then the response is $(1 + \cdots + 7) - (a + b + c + d)$. (Incidentally, this rule holds for all the queries in this question.) Now determine $\mathrm{DK}(i)$, $i = 1, \ldots, 7$.

5. Apply the results of this section to queries of type geometric mean. In particular give analogs to Theorems 10 and 11. Make sure to state your assumptions explicitly. Specifically the necessity of Theorem 10 will hold only under certain assumptions about the database.

Bibliographic Note

The notion of restricted overlap is from Dobkin *et al.*;[28] Theorem 10 is taken from there (sufficiency) and from Reiss[76] (necessity). Theorem 11 is from Leiss;[51] similar results have been independently obtained in Davida *et al.*,[17] using finite projective planes, and in Schwartz,[86] also using symmetric block designs.

2.3.1.3. Selector Functions

In this section we continue our investigation of key-specified queries. While in Sections 2.3.1.1 and 2.3.1.2 we dealt with averages, we will concern ourselves now with selector functions. A selector function for our purposes is a function f of k arguments, say, $x_1, \ldots, x_k$, such that the value of $f(x_1, \ldots, x_k)$ is one of these values $x_1, \ldots, x_k$,

$$f(x_1, \ldots, x_k) \text{ in } \{x_1, \ldots, x_k\}$$

Thus the response of a query $(i_1, \ldots, i_k)$ of this type f will always be an element DK(s) of the database, namely,

$$f(\text{DK}(i_1), \ldots, \text{DK}(i_k))$$

This is in contrast to averages since the average of some number of elements need not be an element of the database. Commonly used selector functions are medians, maxima, and minima. In subsequent parts of this section we will concentrate on these often used queries. First, however, we will show a very surprising result concerning arbitrary selector functions.

Throughout this entire section we will assume without further mentioning that no elements of the database are known to the user, $D_0 = \emptyset$.

2.3.1.3.1. Arbitrary Selector Functions. In this section we will show that it is possible to compromise a database even with arbitrary selector functions. In other words, the functions we consider could be entirely unpredictable; in fact they could be random functions (that is, display behavior entirely outside of the user's influence or knowledge) provided that the result they return is always one of the argument values. How this value is selected from the argument values is immaterial. We will have to make another assumption, namely, that the values of the database are pairwise different,

$$\text{DK}(i) \neq \text{DK}(j) \qquad \text{for } i \neq j$$

Under these conditions we can state the following:

Theorem 12. Assume that the elements of the database are pairwise distinct, assume that there are at least k^2 elements, $N \geq k^2$, and let the queries to be used be of type arbitrary selector function. In at most $k^2 + 1$ queries it is possible to compromise the database. This holds even if the overlap of the queries is restricted to at most 1.

Remark. The compromise in this statement is neither selective nor global. This is one of the very few instances where one can compromise but it is not possible to specify a certain element one wishes to determine. The determined element in this compromise is known only at the end of the construction, as will be seen from the proof.

Proof. The proof is in two steps. First we show the claim based on an assumption and then we prove the assumption. This assumption is as follows: There exist m sets of k indices each,

$$S_1, \ldots, S_m$$

with the following properties:

(i) Any two sets have at most one index in common,

$$\text{card}(S_i \cap S_j) \leq 1 \qquad \text{for all } i \neq j$$

(ii) The set S_i is a subset of $\{1, \ldots, m-1\}$ for all $i = 1, \ldots, m$.

Assuming the existence of such sets we proceed as follows. Let q_i be the response to the query $(j \mid j \text{ in } S_i)$; since any two sets have at most one element in common these queries certainly have overlap at most 1. By assumption that the queries are of type arbitrary selector function we know that

$$q_i \text{ in } \{1, \ldots, m-1\} \qquad \text{for all } i = 1, \ldots, m$$

Note that there are m responses but only $m - 1$ possibilities; thus for some i and j with $i \neq j$ we must have $q_i = q_j$. But since $\text{card}(S_i \cap S_j)$ is at most 1 and since the elements of our database are pairwise distinct, it follows that

$$\text{DK}(x) = q_i = q_j \qquad \text{for } S_i \cap S_j = \{x\}$$

This conclude the proof of the theorem. □

We now have to give the proof of the assumption.

Proposition 4. There exist sets $S_1, \ldots, S_m$ such that (a) S_i is a subset of $\{1, \ldots, m-1\}$ for all $i = 1, \ldots, m$; (b) S_i has exactly k elements, $\text{card}(S_i) = k$, for $i = 1, \ldots, m$; (c) $\text{card}(S_i \cap S_j) \leq 1$ for all $i \neq j$.

Proof. By the theory of symmetric block designs we know (see Corollary 5) that for every $k \geq 2$ there exists a square matrix $E_{k,1}$ of dimension $k(k-1) + 1$ consisting of zeros and ones only such that

(i) every row contains exactly k ones, and (ii) any two rows have exactly one one in the same column. It follows that every column also has exactly k ones. Now consider the matrix $E_{k+1,1}$ of dimension $k(k+1)+1$. Remove one row from $E_{k+1,1}$ and also remove the $k+1$ columns in which this row has ones. The remaining matrix E' has $k(k+1)$ rows and k^2 columns. Furthermore, in this new matrix every row contains exactly k ones as every column had one one in the same position as the row we removed. Now let each row of E' correspond to a set S_i of numbers as follows:

$$S_i = \{j \mid \text{row } i \text{ of } E' \text{ has a one in column } j\}$$

Clearly any two of these sets have at most one element in common,

$$\operatorname{card}(S_i \cap S_j) \leq 1 \qquad \text{for } i \neq j$$

each set has exactly k elements,

$$\operatorname{card}(S_i) = k \qquad \text{for } i = 1, \ldots, k(k+1)$$

and all sets are subsets of $\{1, \ldots, k^2\}$ after appropriate renumbering (each of the k^2 columns corresponds to an element). Since

$$k(k+1) > k^2 \qquad \text{for } k \geq 2$$

the claim follows. □

Thus the proof of Theorem 12 is complete. It should be clear now why selective compromise is not possible; it is impossible to predict which two responses will be equal; therefore it is also impossible to determine a specific element.

Example. We first illustrate the construction in the proof of Proposition 4. Recall the matrix $E_{3,1}$ on p. 64.

$$\begin{bmatrix} 1 & 1 & 1 & 0 & 0 & 0 & 0 \\ 1 & 0 & 0 & 1 & 1 & 0 & 0 \\ 0 & 1 & 0 & 1 & 0 & 1 & 0 \\ 0 & 0 & 1 & 0 & 1 & 1 & 0 \\ 1 & 0 & 0 & 0 & 0 & 1 & 1 \\ 0 & 1 & 0 & 0 & 1 & 0 & 1 \\ 0 & 0 & 1 & 1 & 0 & 0 & 1 \end{bmatrix}$$

Let us remove the last row and all the columns where this row has a one; this yields the following (6, 4) matrix:

$$\begin{bmatrix} 1 & 1 & 0 & 0 \\ 1 & 0 & 1 & 0 \\ 0 & 1 & 0 & 1 \\ 0 & 0 & 1 & 1 \\ 1 & 0 & 0 & 1 \\ 0 & 1 & 1 & 0 \end{bmatrix}$$

Note that each row contains two ones, each column contains three ones, and any two rows have overlap at most one (rows 1 and 4, for example, have no overlap at all). This yields the following sets:

$$\{1, 2\}, \quad \{1, 3\}, \quad \{2, 4\}, \quad \{3, 4\}, \quad \{1, 4\}, \quad \{2, 3\}$$

Any two of these six sets have at most one element in common, and they are all subsets of the set $\{1, 2, 3, 4\}$. These sets can now be used to compromise by posing one query for each of them, i.e.,

$$(1, 2), \quad (1, 3), \quad (2, 4), \quad (3, 4), \quad (1, 4), \quad (2, 3)$$

As our queries are of type selector function it follows that there are exactly *four* possibilities for the responses to these *six* queries. Thus at least one of the responses must be repeated. In fact it suffices to take any five of these six queries to achieve compromise. All we have to do is determine the element which is common to the two queries with the same response; recall that we assumed that all the elements of the database are pairwise distinct.

It is important to note that Theorem 12 analyzes the worst possible case, i.e., where everything goes against the would-be compromiser. The other end of the spectrum where the compromiser is uncommonly lucky is exemplified by the following:

Proposition 5. Assume that all elements are distinct. There are cases where already two queries of type arbitrary selector function permit to compromise the database.

Proof. Consider the two queries $(i_j \mid j = 1, \ldots, k)$ and $(i_j \mid j = k, \ldots, 2k - 1)$. These two queries have precisely i_k in common. Assume now that the responses to the two queries are both y. In this case it follows that $\mathrm{DK}(i_k) = y$ as by assumption all the elements are pairwise distinct. □

These results indicate that selector functions in general define an altogether unpleasant class of queries from the viewpoint of database security as even the weirdest selector functions can be used to compromise a database in $k^2 + 1$ queries. This impression is amply borne out by other results, notably for maxima, minima, and medians. The case of minima, to be dealt with next, is interesting as one can show some trade-offs between the number of queries, the minimum number of elements in the database, and the overlap required to guarantee compromise.

2.3.1.3.2. Maxima and Minima. In this paragraph we consider the case of a database with queries of type maximum. It should be easy to see that any compromise with queries of type maximum is possible iff compromise with queries of type minimum is possible; just define the total order $\leq'$ on numbers by $x \leq' y$ iff $y \leq x$, where $\leq$ is the usual order, and apply the compromising sequence of queries now with minima instead of maxima. These queries are interesting from two aspects: they are fairly common functions, and we can show some trade-offs between the size of the database, the number of queries sufficient to compromise, and the permitted overlap. Throughout the whole paragraph we will assume that all elements of the database are pairwise distinct, i.e., $\mathrm{DK}(i) = \mathrm{DK}(j)$ implies $i = j$.

Theorem 13. Let all elements of the database be distinct. Then the database can be compromised with k queries of type maximum with overlap 1 provided it has at least $k(k+1)/2$ elements:

$$N \geq k(k+1)/2$$

Proof. Consider a (k, k) matrix M where for simplicity we write j instead of i_j. The matrix is symmetric,

$$m_{i,j} = m_{j,i} \qquad \text{for all } i, j \text{ in } \{1, \ldots, k\}$$

and if we string it out row by row taking only the part above and including the main diagonal, we get the consecutive numbers from 1 to $k(k+1)/2$, i.e.,

$$\begin{array}{cccccc}
1 & 2 & 3 & 4 & \cdots & k \\
 & k+1 & k+2 & k+3 & \cdots & 2k-1 \\
 & & 2k & 2k+1 & \cdots & 3k-3 \\
 & & & \cdot & \cdots & \cdot \\
 & & & \cdot & \cdots & \cdot \\
 & & & & \cdots & \cdot \\
 & & & & \cdot & \cdot \\
 & & & & & k(k+1)/2
\end{array}$$

It can be verified that any two queries have at most one index in common. Furthermore note that exactly the elements in the main diagonal of M appear once, all others appear twice. Let each row of the matrix correspond to a query in the usual way. Using these queries the database can be compromised as follows. Let y be the maximum of all the responses, i.e.,

$$y = \max\{q_1, \ldots, q_k\}$$

Clearly we must have

$$y = \mathrm{DK}(s) \qquad \text{for some } s \text{ in } \{1, \ldots, k(k+1)/2\}$$

We claim that it is possible to determine s. Obviously y is the response to query q_i iff DK(s) is involved in this query. Since any index j appears in at most two queries we distinguish two cases. Either there is only one query, say, q_m with response y; then we can determine s as the unique diagonal element occurring in this query q_m. Or there are two queries with response y, say, q_m and q_n. As they have only one element in common (overlap is 1!) this common element must be s (recall that all elements of the database are assumed to be distinct). □

Again the compromise is neither global nor selective.

Example. Let $k = 4$; we assume that at least ten elements exist in the database. Our four queries are as follows:

$$\begin{aligned} &1, 2, 3, 4 &: \ a\\ &2, 5, 6, 7 &: \ b\\ &3, 6, 8, 9 &: \ c\\ &4, 7, 9, 10 &: \ d \end{aligned}$$

Let m be the maximum of the four responses, i.e.,

$$m = \max(a, b, c, d)$$

There are two possibilities: either m occurs once among the responses a, b, c, and d, or m occurs twice. To illustrate the first case, let $a = 1$, $b = 2$, $c = 3$, and $d = 4$. Clearly $m = d = 4$. This implies that the largest element of the ten is involved in the query (4, 7, 9, 10) and in no other query. Since DK(4), DK(7), and DK(9) occur in other queries, it follows conclusively that

$$\mathrm{DK}(10) = 10$$

To illustrate the other possibility (m occurring twice among a, b, c, and d), let $a = b = 2$ and $c = d = 1$; thus $m = 2$. It follows that the largest element of DK(1), . . . , DK(10) is involved in the queries (1, 2, 3, 4) and (2, 5, 6, 7) and in no other query; since only DK(2) satisfies this we conclude that

$$\mathrm{DK}(2) = 2$$

Let us now see what happens if we allow unrestricted overlap. The reader may recall that usually allowing greater overlap reduces the number of queries required to compromise. This holds also for queries of type maxima as is stated in the following:

Theorem 14. Let all the elements be distinct. Then the database can be compromised within $m := \lceil 1 + \log_2 k \rceil$ queries with unlimited overlap provided $N \geq 2^{m-1}$. ($\lceil x \rceil$ denotes the smallest integer at least as large as x.)

Proof. The main idea of the proof is as follows; again we will write j instead of i_j. There are 2^{m-1} nonempty subsets of a set with m elements. By definition each of the m elements will appear in precisely k of the subsets. Number the subsets 1, 2, . . . , 2^{m-1} and define queries $q_1, \ldots, q_m$ as follows: j is involved in query i iff element i is contained in subset j. It follows that there are k elements involved in each of the m queries. Consider the responses to these m queries and determine their maximum, say, y. Then determine precisely all the responses which were equal to y. [Since $y = \mathrm{DK}(s)$ for some s in $\{1, \ldots, 2^{m-1}\}$, at least one response must be y.] Since each of the elements j is uniquely identified by its presence or absence in the m queries, this set of all responses which are equal to y also identifies uniquely one j, which achieves the compromise. □

Example. Let $k = 8$; consequently $m = 4$. We have 15 nonempty subsets of a set with four elements. Consider

element \ subset	1	2	3	4	5	6	7	8	9	10	11	12	13	14	15
1	1	1	1	1	0	1	1	1	0	0	0	1	0	0	0
2	1	1	1	0	1	1	0	0	1	1	0	0	1	0	0
3	1	1	0	1	1	0	1	0	1	0	1	0	0	1	0
4	1	0	1	1	1	0	0	1	0	1	1	0	0	0	1

where we write 1 if element i occurs in subset j; otherwise 0. Clearly every subset is identified by a unique "pattern" of zeros and ones. This translates into the following set of four queries:

q_1 : (1, 2, 3, 4, 6, 7, 8, 12), corresponding to the first line
q_2 : (1, 2, 3, 5, 6, 9, 10, 13), corresponding to the second line
q_3 : (1, 2, 4, 5, 7, 9, 11, 14), corresponding to the third line
q_4 : (1, 3, 4, 5, 8, 10, 11, 15), corresponding to the last line

Suppose now that the maximum y of the responses to these four queries is equal to the responses of the first, third, and fourth query. Therefore the element equal to y is DK(4), as only 4 has a one in the first, third, and fourth line and a zero in the second. Similarly, if y is equal to the response of the third query only, then the element of the database equal to y is DK(14).

Theorem 15. Let all the elements of the database be distinct, and let the queries be of type maximum. Then the minimum number of elements of a database where compromise can be guaranteed is $k + 1$, and in this case exactly k queries of unrestricted overlap are required.

Proof. It is clear that a database with fewer than $k + 1$ elements cannot be compromised as there is at most one query which then would involve all the elements. Thus we merely have to show that compromise is possible with $k + 1$ elements. Choose any $k + 1$ elements and determine k arbitrary subsets of k elements; clearly there are precisely $k + 1$ such subsets. Each subset is characterized by its bit pattern; this is a string of $k + 1$ zeros and ones with a one in position i iff element i does not appear in this subset. Form a matrix out of these bit patterns treating each one as a row. This yields a $(k, k + 1)$ matrix with the property that any two columns are distinct. This can be seen as follows: If we did this for all $k + 1$ subsets of k elements we could reorder the $k + 1$ rows in such a way that the resulting matrix is a $(k + 1, k + 1)$ unit matrix. Removing one row will yield one column with zeros only and the other columns all have a one but each in a different position. This matrix is exactly the matrix we are dealing with. It has the property that each column uniquely identifies a different maximum (as in the proof of Theorem 14 and in the example). This proves sufficiency. To see the necessity of k queries observe that one query less is equivalent to removing another row. This results in a $(k + 1, k - 1)$ matrix which contains two equal columns, namely, two columns of zeros only. Therefore it is possible that we know the maximum is in one of the two sets of indices but we cannot decide in which one without asking another query. □

Example. Let us illustrate how to use the proof of Theorem 15 in order to attain compromise. Let $k = 5$; consequently our database must have at least six elements. The matrix corresponding to *all* subsets of five of these six elements is

$$\begin{bmatrix} 1 & 0 & 0 & 0 & 0 & 0 \\ 0 & 1 & 0 & 0 & 0 & 0 \\ 0 & 0 & 1 & 0 & 0 & 0 \\ 0 & 0 & 0 & 1 & 0 & 0 \\ 0 & 0 & 0 & 0 & 1 & 0 \\ 0 & 0 & 0 & 0 & 0 & 1 \end{bmatrix}$$

where each row corresponds to a query, e.g., the second row to the query (1, 3, 4, 5, 6). Now remove one row, for instance the first:

$$\begin{bmatrix} 0 & 1 & 0 & 0 & 0 & 0 \\ 0 & 0 & 1 & 0 & 0 & 0 \\ 0 & 0 & 0 & 1 & 0 & 0 \\ 0 & 0 & 0 & 0 & 1 & 0 \\ 0 & 0 & 0 & 0 & 0 & 1 \end{bmatrix}$$

Clearly any two columns are distinct. This matrix corresponds to the queries

$$\begin{aligned} 1, 3, 4, 5, 6 \quad &: \quad a \\ 1, 2, 4, 5, 6 \quad &: \quad b \\ 1, 2, 3, 5, 6 \quad &: \quad c \\ 1, 2, 3, 4, 6 \quad &: \quad d \\ 1, 2, 3, 4, 5 \quad &: \quad e \end{aligned}$$

The first possibility is that all five responses are the same. As all elements are assumed to be pairwise distinct this implies that the element involved in all queries has this value, i.e., DK(1). Otherwise, as every element is involved in at least four queries, four responses must be the same and the fifth is smaller. Thus if, for example, $a = b = c = d > e$, it is clear that DK(6) $= a$, or if $a = c = d = e > b$ then DK(3) $= a$.

These three theorems indicate interesting trade-offs between various parameters of a database related to its security. Note that Theorem 15 does not state that k queries are always necessary (this would be wrong by Proposition 5); it merely states that even in the most unfavorable situation for the compromiser k queries suffice.

2.3.1.3.3. Medians. In this section we will primarily summarize and discuss results which have been obtained for queries of type median; in view of Theorem 13 we refer to the original papers for most of the proofs.

For the sake of completeness let us state what the median of a set of numbers is. Given k numbers $y_1, \ldots, y_k$ their median is the $k/2$ largest of these numbers if k is even and the $(k+1)/2$ largest of them if k is odd. Let us first assume that all elements of the database are pairwise distinct. In this case Proposition 5 applies and states that two queries are sometimes sufficient for compromise. In any case we have (for simplicity let k be odd) the following:

Theorem 16. Let all elements be distinct and let the queries be of type median. At most $3(k+1)/2+2$ queries are sufficient for compromise provided that the database has at least $k+2$ elements:

$$N \geq k+2$$

Proof. Let $p=(k+1)/2$ and let $\mathrm{DK}(i_1), \ldots, \mathrm{DK}(i_{k+2})$ be $k+2$ distinct data elements. There are $k+1$ possible medians we can determine using the first $k+1$ elements ($k+1$ different subsets of k elements each). The corresponding medians for each of them will be one of only two possible answers, g or h; WLOG $g<h$. Define

$$G=\{\mathrm{DK}(i_j) \mid \mathrm{DK}(i_j) \leq g\} \quad \text{and} \quad H=\{\mathrm{DK}(i_j) \mid \mathrm{DK}(i_j) \geq h\}$$

Clearly $\mathrm{DK}(i_j)$ is in G (in H) iff the median of the set $\{\mathrm{DK}(i_1), \ldots, \mathrm{DK}(i_{k+1})\} - \{\mathrm{DK}(i_j)\}$ is equal to h (to g). One can verify that G and H have exactly p elements. Now form G' by removing any two elements from G and determine the median m of

$$G' \cup H \cup \{\mathrm{DK}(i_{k+2})\}$$

Since by definition of G' this response cannot be less than h there are two possibilities: the response is equal to h, which implies $\mathrm{DK}(i_{k+2})<h$, or the response is greater than h, in which case $\mathrm{DK}(i_{k+2})$ is also greater than h. Therefore at this point we know whether $\mathrm{DK}(i_{k+2})$ is greater or

less than h. [The case $\mathrm{DK}(i_{k+2}) = h$ is not possible as the elements are distinct and h is the response to a query which does not involve $\mathrm{DK}(i_{k+2})$.] Note that we used $k + 2$ queries to decide this. WLOG $\mathrm{DK}(i_{k+2}) < h$. Let H_0 be a set of $p - 1$ elements of H and determine the median of $(G \cup \{\mathrm{DK}(i_{k+2})\} \cup H_0) - \{\mathrm{DK}(i_j)\}$ for all $\mathrm{DK}(i_j)$ in $G \cup \{\mathrm{DK}(i_{k+2})\}$. This adds another $p + 1$ queries for a total of $3(k + 1)/2 + 2$ queries. Now it can be seen that one median, say, y, occurs p times and one median occurs once. This implies $y = \mathrm{DK}(i_j)$, where $\mathrm{DK}(i_j)$ is not in the set which has the median different from y. □

Example. Let $k = 5$; we will show that 11 queries of type median are sufficient for compromising provided the database has at least seven elements and the elements DK(1), ..., DK(7) are pairwise distinct. It is clear that we can form exactly six different queries using the elements DK(1) through DK(6), and furthermore it follows that only two different responses are possible to these queries:

1, 2, 3, 4, 5 : 4
1, 2, 3, 4, 6 : 3
1, 2, 3, 5, 6 : 3
1, 2, 4, 5, 6 : 3
1, 3, 4, 5, 6 : 4
2, 3, 4, 5, 6 : 4

Consequently the sets G and H are as follows:

$$G = \{1, 2, 6\} \quad \text{and} \quad H = \{3, 4, 5\}$$

Recall that DK(6) is in G because the median of DK(1), DK(2), DK(3), DK(4), and DK(5), i.e., the response to the first query, is 4; the other elements follow similarly. Now let us remove the elements 1 and 2 from G and pose the query

(3, 4, 5, 6, 7)

let the response be 4. Thus we know that DK(7) < 4. Let the set H_0 be {3, 4}; we pose the following queries:

2, 3, 4, 6, 7 : 3
1, 3, 4, 6, 7 : 2
1, 2, 3, 4, 7 : 3
1, 2, 3, 4, 6 : 3

It is clear that $y = 3$ and therefore

$$\mathrm{DK}(2) = 3$$

For completeness we give the values of all seven elements:

i	DK(i)
1	0
2	3
3	4
4	6
5	5
6	2
7	1

This gives an upper bound of $O(k)$ on the number of queries sufficient to guarantee compromise with all elements of the database distinct. In fact it can be shown (Dobkin *et al.*[29]) that $O(\mathrm{sqrt}(k))$ queries suffice under these circumstances. Using a probabilistic approach this result can be improved to $O(\log(k)^2)$ queries (Reiss[75]). These results concern simple compromise. While selective compromise is provably impossible for arbitrary selector functions, for medians it can be achieved, however, at some expense. In Dobkin *et al.*[29] it was shown (again for databases with distinct values) that $(d(1 - d))^{k/2}$ queries are sufficient on the average to guarantee selective compromise provided the element y to be determined lies in the d-tile of the database, i.e., at least $d \cdot N$ elements are smaller than y and at least $d \cdot N$ elements are larger than y:

$$0 < d < 1/2$$

This exponential bound has been improved in Reiss[75]; in that method for any d, $0 < d < 1/2$, and sufficiently large N only $O(k)$ queries suffice on the average to determine a particular element in the d-tile.

It can easily be seen that queries of type median cannot be used to determine arbitrary database elements. For instance, it should be obvious that the $k/2 - 1$ largest and the $k/2 - 1$ smallest elements of a database can never be determined. The proof is by contradiction; assume we can in fact determine one of these elements. Then consider another database which is the same as the previous one but we changed all the

$k/2 - 1$ largest elements by adding 1 and the $k/2 - 1$ smallest elements by subtracting 1. Clearly the responses to our compromising sequence of queries remain unchanged. This consequently yields a contradiction. It should be noted that this argument depends crucially on the assumption that all elements of the database are distinct.

These results should be compared with Theorem 12, which contains Theorem 16 as a special case.

In Reiss[76] some results are given for the case where the elements of the database are not all pairwise different. In particular we have an analog to Proposition 4 for medians:

Proposition 6. Assume $N \geq k + 2$. It is possible in some cases to compromise with three queries but there are no cases where fewer queries achieve compromise.

Proof. Consider the queries

$$
\begin{array}{ll}
(i_1, i_2, \ldots, i_k) & \text{with response } a \\
(i_{k+1}, i_2, \ldots, i_k) & \text{with response } b \\
(i_{k+2}, i_2, \ldots, i_k) & \text{with response } c
\end{array}
$$

and assume that

$$b < a < c$$

The first and second queries imply that

$$\mathrm{DK}(i_{k+1}) < a \leq \mathrm{DK}(i_1)$$

the first and third queries imply that

$$\mathrm{DK}(i_{k+2}) > a \geq \mathrm{DK}(i_1)$$

But now we have

$$a \geq \mathrm{DK}(i_1) \geq a$$

which of course implies

$$\mathrm{DK}(i_1) = a$$

Let us now show that no fewer queries suffice. Clearly one query is not enough; so let us consider two queries

$q_1 \ : \ (i_1, \ldots, i_j, i_{j+1}, \ldots, i_k)$ with response a

$q_2 \ : \ (i_{k+1}, \ldots, i_{k+j}, i_{j+1}, \ldots, i_k)$ with response b,

for some j with $1 \leq j \leq k$

Clearly if $a = b$ nothing can be deduced. Thus WLOG $a > b$. This implies that the number of elements in $\{DK(i_1), \ldots, DK(i_j)\}$ which are not less than a is larger than the number of such elements in $\{DK(i_{k+1}), \ldots, DK(i_{k+j})\}$. This conveys the most information if $j = 1$. But even in this case we can only deduce $DK(i_1) \geq a$ or $DK(i_{k+1}) < a$. Therefore at least three queries are needed to determine the value of one data item. □

Reiss then proceeds to show that with this model (i.e., the elements of the database are not necessarily distinct) if one insists on an overlap of at most 1, the number of queries which is sufficient for compromise in the best possible case is at least $3(k + 1)/4$ (for $k \geq 3$) and at most $3k - 5$ provided $N \geq k^2 - 2k + 4$. This result is therefore an analog to Proposition 6 but with overlap at most 1. This should be contrasted with Proposition 5, which incidentally had also overlap at most 1 but where we assumed that all data elements are distinct.

Exercises

1. Compromise a database with queries of type arbitrary selector function of six arguments. More specifically, list all queries required to achieve compromise.

2. Write a program which takes as input k and computes a matrix $E_{k,1}$.

3. Write a program which takes as argument k and a selector function of k arguments (you must find some way of passing a function as argument, e.g., call by name in Algol60, SNOBOL4) and outputs some index i and the corresponding entry $DK(i)$ of a given database. Use the program in Exercise 2 as a subroutine. Keep in mind that all database entries are to be pairwise distinct. Be sure to stop posing queries as soon as two queries respond with the same value.

4. In the example of a database following Theorem 16, assume you do not know DK(7) but the response to query (3, 4, 5, 6, 7) is 5. Describe precisely what queries achieve compromise and explain why.

5. Let k be odd and consider the $(k+1)/2$ smallest and $(k+1)/2$ largest element of a database where all elements are pairwise different. Is it possible by using queries of type median to compromise these two elements selectively (a) in all cases? (b) in special cases? Note that a query involving all $(k+1)/2$ smallest elements of a database and $(k-1)/2$ other elements will respond with the $(k+1)/2$ smallest element, but can this element be found?

6. Can you show an analog to Proposition 6 for maxima and minima? More specifically, can you show that there is a constant c such that it is possible in *some* cases to compromise with c queries of type maximum (minimum) but there are no cases where fewer than c queries suffice? Note that we do *not* assume that all database elements are pairwise distinct.

7. Same question as Exercise 6 but for queries of type arbitrary selector function.

8. Suppose the database elements are not pairwise distinct. Assume that the queries are of type maximum (minimum). Show that it is always possible to compromise this database provided there are sufficiently many elements in the database and there are at least $k+1$ different values. Hint: Ask sufficiently many queries until you get at least $k+1$ different responses. Then apply the usual method to the appropriate elements.

9. Prove formally that in the situation of Exercise 8 it is not possible to compromise if there are fewer than $k+1$ different values.

10. Can you show a statement analogous to Exercises 8 and 9 for queries of type median?

11. Same question as Exercise 10 but for queries of type arbitrary selector function.

Bibliographic Note

Theorem 12 is due to Demillo *et al.*;[19] however, their proof uses projective planes. This implies that either k be a prime power or the construction must be embedded in a larger projective plane. Our direct proof uses symmetric block designs; Proposition 4 is implicitly contained in Leiss.[51] Theorems 13 and 14 are from Davida *et al.*[17] A discussion of the case where the elements of the database are not pairwise distinct can also be found there. In Demillo *et al.*[19] the case is made that the assumption that all elements be distinct is often satisfied in practice when dealing with a relatively small part of the database. Theorem 16 is from Dobkin *et al.*[28] The proof that $O(\text{sqrt}(k))$ queries suffice can be found in Dobkin *et al.*[29] The improvement of this algorithm to $O(\log(k))$ queries is contained in Reiss.[75] Partial selective compromise is considered in Dobkin *et al.*[29] and improved in Reiss.[75] Proposition 6 is from Reiss[76] as is its analog for restricted overlap.

2.3.2. Secure Databases

In the last sections we collected evidence that it is practically impossible to obtain secure databases when using the traditional approach. However, it should be quite obvious that there is a definite and indeed a growing need for secure databases. The more confidential information is stored concerning individuals as well as institutions, the more is it in the interest of these groups to extract the guarantee from the collecting agency that the information is inaccessible to unauthorized persons or institutions, that is, that it cannot be abused. Such a guarantee is impossible if the database used resembles any of those we discussed on the previous pages. Unfortunately almost all of the presently used databases do, as our models are quite realistic. Thus for data collecting agencies, a database which can be shown to be secure becomes almost mandatory. In the following we will describe just that: A database which is almost identical to the conventional databases and yet it can be shown to be secure. Moreover, the method is conceptually very simple, it can be superimposed on practically any existing database, and the additional cost is negligible. In Section 2.3.2.1 we will show how one can implement this method, which we call randomizing, for queries of type average. Section 2.3.2.2 is devoted to implementing randomizing in the case of certain selector functions. In Section 2.3.2.3 we discuss the price we have to pay for this guarantee of security, an efficient and interesting way how to reduce this price, and a trade-off between security and price.

2.3.2.1. Averages

All previous results suggest that restrictions imposed on the form of the queries like overlap (syntactic restrictions) are not very feasible. On the other hand it should be obvious that there is a need for some mechanism to protect the security of a database. In order to achieve this we propose a method using random selection. Rather than using queries of type average of k elements we suggest the following type of queries where $v \geq 0$: The user provides as usual k indices $i_1, \ldots, i_k$ and the system determines $\mathrm{DK}(i_1), \ldots, \mathrm{DK}(i_k)$; but instead of computing the average of these k values the system first determines randomly another v elements of the database and *then* computes the average of these $k + v$ elements. Clearly, for $v = 0$ we obtain the original queries of type average. This method will be called randomizing and the queries will be referred to as randomized queries of type average. Thus if q is the response to the randomized query of type average $(i_1, \ldots, i_k)$, then we have

$$q = \left[\sum_{j=1}^{k} \mathrm{DK}(i_j) + \sum_{j=1}^{v} s_j\right] \Big/ (k + v)$$

where s_j is a result of the random selector function S. In the following we will concentrate on the case where $v = 1$. However, all the results can also be applied to the case where $v > 1$.

For example, if $k = 5$, $v = 2$, and $\mathrm{DK}(i)$ is defined to be $10i$, then the query

$$(9, 15, 30, 60, 137)$$

is computed as follows:

$$(\mathrm{DK}(9) + \mathrm{DK}(15) + \mathrm{DK}(30) + \mathrm{DK}(60) + \mathrm{DK}(137) + s_1 + s_2)/7$$

where s_1 and s_2 are the randomly selected elements, e.g., 870 and 190; in this case the randomized response is 510 while the (true) average of the five elements specified in the query is 502. It follows that in this case the relative error of the randomized response is approximately 1.6%. We will discuss the problem of errors later in more detail.

We claim that it is impossible to compromise a database with randomized queries of type average even for very large k. If k is not too small the response of such a query will be a good approximation on the (desired) precise value (average of $\{DK(i_1), \ldots, DK(i_k)\}$). This question of precision will be discussed extensively in Section 2.3.2.3.

For the following we assume that no data elements are known to the user,

$$D_0 = \emptyset$$

Furthermore, let us assume first that our way to compromise a database is by solving a system of linear equations obtained directly from a sequence of queries. We claim that within our framework and under this assumption a database cannot be compromised.

To substantiate our claim we first observe that our best "guess" for the values s_j returned by S is the average

$$s^* = (DK(1) + \cdots + DK(N))/N$$

taken over the whole database or (if we do not know s^*) the average over all responses to the queries issued. Therefore, rather than dealing with a system of equations

$$D^*x = k^*q$$

where D is the matrix derived from our given sequence of queries, we must solve the modified system

$$D^*x = q'$$

where

$$q' = ((k+1)q_1 - s_1, \ldots, (k+1)q_t - s_t)$$

However, all that we can solve is the system

$$D^*x = q''$$

where

$$q'' = ((k+1)q_1 - s^*, \ldots, (k+1)q_t - s^*)$$

Let us assume that we can solve the equation $D^*x = q'$, i.e., D is nonsingular. Another way of looking at this problem is to assume that q' contains errors and is given by q''. The sensitivity of a system of linear equations $M^*x = c$ to errors in c is usually measured by the condition number

$$\text{cond}(M) = || M || * || M' ||$$

where $|| \; ||$ is some matrix norm and M' denotes the inverse matrix to M. We choose the following norm $|| \; ||$:

$$|| M || = \left(\sum_{i,j=1}^{n} | m_{i,j} |^2 \right)^{1/2}, \qquad \text{where } M = (m_{i,j}) \qquad (1 \leq i, j \leq n)$$

Let us now determine a lower bound on cond(D). Clearly,

$$|| D || = \text{sqrt}(k^*t)$$

with t being the dimension of D. We claim that

$$|| D' || \geq \text{sqrt}(t/k)$$

where D' is the inverse matrix to D. Let $D = (d_{i,j})$, $D' = (d'_{i,j})$. Since D^*D' is the unit matrix we have

$$d_{i,1}d'_{1,i} + \cdots + d_{i,t}{}^*d'_{t,i} = 1$$

We want to obtain a lower bound on

$$\left(\sum_{j=1}^{t} | d'_{j,i} |^2 \right)^{1/2}$$

for all $i = 1, \ldots, t$. So let $d'_{j,i} = 0$ whenever $d_{i,j} = 0$. Therefore at most k of the $d'_{j,i}$ are nonzero. Furthermore these must add up to 1; note that $d_{i,j}$ is either 0 or 1. Now consider

$$\left(\sum_{j=1}^{k} | z_j |^2 \right)^{1/2}$$

as a function in the k variables $z_1, \ldots, z_k$ and determine a minimum subject to the constraint $z_1 + \cdots + z_k = 1$. This minimum is attained for

$$z_1 = \cdots = z_k = 1/k$$

and is $1/\text{sqrt}(k)$. This holds for all $i = 1, \ldots, t$; hence we get

$$|| D' || \geq \text{sqrt}(t/k)$$

and therefore

$$\text{cond}(D) \geq t$$

This implies that all matrices D are quite sensitive to errors; note that $t > k$. Furthermore the errors will be considerable since on the average $s_j - s^*$ will be of the same order of magnitude as s^*. Therefore it is not possible to obtain useful results when solving this system of equations; the database cannot be compromised.

Example. The most convenient matrix which can be used for compromising with queries of type average where the overlap is unrestricted is that discussed in Section 2.3.1.1 given by $B = (b_{ij})$ for $1 \leq i, j \leq k + 1$ with $b_{ij} = 0$ for $i = j$ and $b_{ij} = 1$ otherwise. It is also the smallest of all nonsingular matrices which guarantee compromise. Let us determine the condition number of B. First we derive the inverse B' of B. Using the method for solving systems of equations involving B which is given in the proof of Theorem 7, one deduces that $B' = (b'_{ij})$, where

$$b'_{ij} = -(k - 1)/k \text{ if } i = j \qquad \text{and} \qquad b'_{ij} = 1/k \text{ otherwise}$$

The easiest verification of this claim follows by multiplying B and B'; clearly this yields the identity matrix. Now we can determine the condition number of B:

$$\begin{aligned}\text{cond}(B) &= \left(\sum_{i,j=1}^{k+1} |\, b_{ij} \,|^2\right)^{1/2} * \left(\sum_{i,j=1}^{k+1} |\, b'_{ij} \,|^2\right)^{1/2} \\ &= \text{sqrt}(k(k + 1))*\text{sqrt}((k^2 - k + 1)(k + 1)/k^2) \\ &= (k + 1)\text{sqrt}(k - 1 + 1/k)\end{aligned}$$

or

$$\text{cond}(B) = O(k\,\text{sqrt}(k))$$

We list cond(B) for some value of k:

k	Cond(B)
5	>12
10	>33
20	>91
50	>357
100	>1004

We illustrate the effect of the condition number with a detailed example. Let $k = 5$, $v = 1$, and assume that the DK(i) are as follows:

i	DK(i)
1	30
2	150
3	300
4	−50
5	140
6	200

Thus we get the following system:

$$\left[\begin{array}{cccccc|c} 0 & 1 & 1 & 1 & 1 & 1 & 148 \\ 1 & 0 & 1 & 1 & 1 & 1 & 124 \\ 1 & 1 & 0 & 1 & 1 & 1 & 94 \\ 1 & 1 & 1 & 0 & 1 & 1 & 164 \\ 1 & 1 & 1 & 1 & 0 & 1 & 126 \\ 1 & 1 & 1 & 1 & 1 & 0 & 114 \end{array}\right] \begin{array}{c} q_1 \\ q_2 \\ q_3 \\ q_4 \\ q_5 \\ q_6 \end{array}$$

However, as we randomize, these responses are changed as follows:

i	DK(s_i)	q_i
1	150	148.333
2	120	123.333
3	90	93.333
4	180	166.667
5	130	126.667
6	110	113.333

Comparing the actual responses (randomized) with the true averages of the elements involved in the queries reveals that the relative errors range between 1.7% and 0.2%, i.e., are quite negligible as far as precision of the responses is concerned. In fact, the average error is only 0.7%. However, if we compute the DK(i) from the randomized responses we obtain the following values:

i	DK(i)
1	10.28
2	160.28
3	340.28
4	–99.72
5	140.28
6	220.28

Consequently we have the following relative errors:

i	Error
1	65.7%
2	6.9%
3	13.4%
4	99.4%
5	0.2%
6	10.1%

The average error of the computed responses is therefore 32.6%. This substantial increase in the error must be attributed to the size of the condition number. One can see from this example and the table of the condition numbers that for larger k the results computed from the randomized responses are quite meaningless.

Let us now show that the method of "filtering out" the "noise" introduced by the "errors" does not achieve compromise either. This method can be described as follows: Let

$$q_j : (i_{j,1}, \ldots, i_{j,k}) \qquad \text{for } j = 1, \ldots, t$$

be a sequence of queries of type randomized average corresponding to a matrix D of dimension t such that the following holds: If q_j' is q_j as a query of type (nonrandomized) average, then by solving $D^*x = k^*q'$ we can determine all DK(i_1), ... , DK(i_t). By Section 2.3.1.1 such a sequence exists. Now define $q_{j,m}$ to be the response to the mth repetition of query q_j; note that in general $q_{j,m} \neq q_{j,n}$ since the result s_m of the random selector function S in the mth repetition will usually differ from its result s_n in the nth repetition. However,

$$(q_{j,1} + \cdots + q_{j,N^*})/N^*$$

might converge to a certain value with increasing N^*, say, to q_j^*; thus we would get

$$q_j^* = \sum_{n=1}^{t} d_{j,n} \cdot \mathrm{DK}(i_n) + s_j^*$$

where $D = (d_{j,n})$. Assume this holds for all $j = 1, \ldots, t$. Furthermore, assume that

$$s_1^* = \cdots = s_t^* = s^*$$

i.e., the value to which S converges does not depend upon the particular query it is used in. We will assume that s^* is known; usually it can be obtained by averaging. Under these circumstances it is possible to compromise the database since this simply amounts to solving the following system of linear equations:

$$D^* x = p^*$$

where

$$p^* = (q_1^* - s^*, \ldots, q_t^* - q^*)$$

This suggests several methods to prevent a breach of security. One is to do away with the assumption that s^* be known; one might for instance change it from time to time. Another is to make sure that N^* is very large, preferably $N^* = O(N)$. A third possibility is to drop the assumption $s_1^* = \cdots = s_t^*$. The first possibility is rather simpleminded; to change s^* from time to time will introduce a time dependency which might be undesirable for reasons outside of the scope of the question we deal with here. We also discard the second strategy because in order to implement it queries must be counted. This would lead to problems similar to those in connection with restricted overlap such as the problem of storing information about queries for a long period of time or the problem of users combining their results. We therefore consider the third proposal, namely, dropping the assumption

$$s_1^* = \cdots = s_t^*$$

By applying the arguments used above it is not difficult to see that the database cannot be compromised if there is a possibility that $s_j^* \neq s_n^*$

for $j \neq n$. Of course we assume that no s_j^* is known; note that this information cannot be retrieved from the queries if one uses the following scheme. Let T be a uniform random selector function; we do not assume anything else about T. Whenever a query of type randomized average is posed, *two* calls to T are made yielding t_1 and t_2. We determine which one of these two values is to be returned as value S in such a way that it "depends on all $\mathrm{DK}(i_j)$ for $j = 1, \ldots, k$". By this we mean the following: If $(i_1, \ldots, i_k)$ is the sequence of indices specified by the user in the query and the choice of S based on this sequence is $\max\{t_1, t_2\}$ (is $\min\{t_1, t_2\}$), then for all i_j, $j = 1, \ldots, k$, there exists an index g_j in $\{1, \ldots, N\}$ such that the choice of S based on the new sequence of indices

$$(i_1, \ldots, i_{j-1}, g_j, i_{j+1}, \ldots, i_k)$$

is $\min\{t_1, t_2\}$ (is $\max\{t_1, t_2\}$). A concrete scheme to achieve this is the following. Let E be the Boolean expression

$$E = [(\mathrm{DK}(i_1) \leq \mathrm{DK}(i_2)) \oplus \cdots \oplus (\mathrm{DK}(i_{k-1}) \leq \mathrm{DK}(i_k))]$$

where $\oplus$ is the exclusive-or operator defined by

$\oplus$	True	False
True	False	True
False	True	False

Then S yields the value $\max\{t_1, t_2\}$ iff E is True. This scheme satisfies the above condition and has the additional advantage that on the average $\max\{t_1, t_2\}$ and $\min\{t_1, t_2\}$ are returned equally often. Note that in this way no information about the range of the $\mathrm{DK}(s)$ is required, no time dependency is introduced, and no previous values of T must be stored. Furthermore it can be implemented at virtually no additional cost. We do realize that a slight bias in the statistical information might be introduced by this method but we feel that this is negligible. We summarize the results in the following:

Theorem 17. Assume that no elements of the database are known. If only queries of type randomized average are allowed then the database is secure.

Example. Assume $k = 5$ and let us pose the following six queries:

$$(2, 3, 4, 5, 6)$$
$$(1, 3, 4, 5, 6)$$
$$(1, 2, 4, 5, 6)$$
$$(1, 2, 3, 5, 6)$$
$$(1, 2, 3, 4, 6)$$
$$(1, 2, 3, 4, 5)$$

For each query the method requires $2v$ calls to the random selector function. Let us assume $v = 1$. (It should be clear that the influence of the scheme becomes more pronounced with larger v!) Let the values of DK(i) be as follows:

i	DK(i)
1	2
2	6
3	5
4	4
5	3
6	3

We determine the Boolean expression E for each query:

$$\mathrm{DK}(2) \underset{F}{\leq} \mathrm{DK}(3) \underset{F}{\leq} \mathrm{DK}(4) \underset{F}{\leq} \mathrm{DK}(5) \underset{T}{\leq} \mathrm{DK}(6) \quad :: \quad T$$

$$\mathrm{DK}(1) \underset{T}{\leq} \mathrm{DK}(3) \underset{F}{\leq} \mathrm{DK}(4) \underset{F}{\leq} \mathrm{DK}(5) \underset{T}{\leq} \mathrm{DK}(6) \quad :: \quad F$$

$$\mathrm{DK}(1) \underset{T}{\leq} \mathrm{DK}(2) \underset{F}{\leq} \mathrm{DK}(4) \underset{F}{\leq} \mathrm{DK}(5) \underset{T}{\leq} \mathrm{DK}(6) \quad :: \quad F$$

$$\mathrm{DK}(1) \underset{T}{\leq} \mathrm{DK}(2) \underset{F}{\leq} \mathrm{DK}(3) \underset{F}{\leq} \mathrm{DK}(5) \underset{T}{\leq} \mathrm{DK}(6) \quad :: \quad F$$

$$\mathrm{DK}(1) \underset{T}{\leq} \mathrm{DK}(2) \underset{F}{\leq} \mathrm{DK}(3) \underset{F}{\leq} \mathrm{DK}(4) \underset{F}{\leq} \mathrm{DK}(6) \quad :: \quad T$$

$$\mathrm{DK}(1) \underset{T}{\leq} \mathrm{DK}(2) \underset{F}{\leq} \mathrm{DK}(3) \underset{F}{\leq} \mathrm{DK}(4) \underset{F}{\leq} \mathrm{DK}(5) \quad :: \quad F$$

It follows that for the first and fifth queries the larger of the two values for the randomly selected value is to be chosen, whereas for the remaining queries the smaller value is to be used. Clearly this scheme introduces a certain bias; however, it ensures that equivalent queries, i.e., queries which involve the same indices as, for instance, (1, 2, 3, 4, 5) and (5, 1, 3, 4, 2), *may* or may *not* be biased in the same way. Moreover, whether this is the case cannot be determined by the user. Note that this even holds if the user is able to create an entry which is larger or smaller than any other database element; this is due to the properties of the exclusive-or operator. Consider, for instance, the three equivalent queries

$$(1, 3, 4, 5, 6)$$
$$(5, 1, 6, 4, 3)$$
$$(6, 5, 4, 3, 1)$$

We know already that the Boolean expression E for the first query is F, for the second query we get $F \oplus T \oplus T \oplus F = F$, and for the third we have $T \oplus T \oplus T \oplus F = T$. Note, however, that it is not correct to conclude that reversing the order of the indices always gives T (F) if the original query gives F (T), since $x \leq x$ and $x \geq x$ are both true.

Leiss[57] reports on a simulation which compares the simpleminded method of selecting the random element (see p. 88ff.) with the more complicated way just outlined. The results of this simulation conclusively demonstrate that a database where the random element DK(s) is selected in a uniformly random way cannot be considered secure. The following table summarizes the results; it lists the probability that the computed value of a database element (compromised element) is afflicted with an error of less than 8%. There are four columns; the first column reports the probability in percent if the query was posed once, while the other three columns report this probability if the user attempted to filter out the error with 10, 100, and 1000 repetitions, respectively, of the queries used in this compromise:

	Repetitions			
k	1	10	100	1000
5	12.42	36.17	53.54	61.11
10	9.43	35.14	68.17	74.09
20	8.84	34.39	70.79	81.52
50	8.55	33.37	71.54	88.70
100	8.17	32.78	71.80	90.42

Probability in % that error of compromised database element is less than 8% (simple method)

On the other hand, if one applies the more sophisticated method using the logical formula E and choosing the larger of two elements (both randomly selected) if E is true and the smaller if E is false, the following table is obtained:

	Repetitions			
k	1	10	100	1000
5	12.45	22.79	25.98	24.55
10	10.14	13.92	9.90	9.27
20	8.91	9.44	3.27	1.58
50	8.45	6.72	0.37	0.02
100	8.14	6.03	0.06	0.00

Probability in % that error of computed database element is less than 8% (sophisticated method)

These data clearly indicate that it is not possible to filter out the errors by repeating the queries; in fact, the errors may even become larger. For more detail see the original paper.

Let us now assume that we know some of the elements in the database,

$$D_0 \neq \emptyset$$

As already pointed out this is far more likely a situation and any method for guaranteeing security will have to be tested at this problem. We do feel that the method of randomizing is very useful partly on the basis of its behavior in the case of D_0 nonempty. In fact we claim that even if we known $k-1$ elements, i.e., all but one of the elements we can specify in a query, it is difficult to determine the remaining single element. It should be noted that there is no difference between knowing $k-1$ elements and knowing arbitrarily many elements, as clearly in a query where a user can specify k indices, one element should be unknown if the user is to learn anything new from the response. Also note that posing a query where all k elements are known will not reveal much about the way the random selection is performed; the user is able to determine the value of the selected database element in this case (provided $v = 1$), but the user is *not* able to determine what index this value has in the database.

Let $\text{DK}(i_1), \ldots, \text{DK}(i_{k-1})$ be the known elements and let $\text{DK}(i_k)$ be the unknown element. Therefore the result q_n of a query $(i_1, \ldots,$

$i_{k-1}, i_k)$ will be

$$q_n = (X + \mathrm{DK}(i_k) + s_n)/(k+1)$$

where

$$X = \mathrm{DK}(i_1) + \cdots + \mathrm{DK}(i_{k-1})$$

note that X is known. Repeating this query will not give the same result since the s_n may change but the sequence

$$(s_1 + s_2 + \cdots + s_m)/m$$

converges to a number s^* and hence the sequence

$$(q_1 + q_2 + \cdots + q_m)/m$$

will also eventually converge to a number q^*. (This may be easily verified using infinite series.) One observes that the rate of convergence and consequently the accuracy with which one can determine $\mathrm{DK}(i_k)$ depends solely on the behavior of the random selector function S. Provided that a function T has been chosen for which

$$(t_1 + \cdots + t_m)/m$$

converges slowly the database can be considered safe for practical purposes (not, however, theoretically).

Let us now assume that we know h elements where

$$1 \leq h \leq k-2$$

In this case the database cannot be compromised. The reason is contained in the fact that this is equivalent to dealing with a database where queries involve $k' = k - h \geq 2$ indices; by the preceding we know that in this case the database cannot be compromised with queries of type randomized average.

Example. Let $k = 5$ and $h = 3$. In this case the obvious queries are of the form

$$(m_1, m_2, m_3; i_1, i_2)$$

where $\mathrm{DK}(m_1)$, $\mathrm{DK}(m_2)$, and $\mathrm{DK}(m_3)$ are the known elements and i_1 and i_2 take the values which enable compromise in the case $k' = k - h = 2$, e.g.,

$$(i_1, i_2) = (1, 2),\ (2, 3),\ (1, 3)$$

(assuming that $\{m_1, m_2, m_3\}$ and $\{1, 2, 3\}$ have no element in common). As we have already shown that compromise is not possible for queries with $k \geq 2$, the claim follows directly. We summarize as follows:

Theorem 18. Suppose some of the elements of the database are known to the user. If $k - 1$ elements or more are known then the database can be compromised with randomized queries of type average; however, a considerable number of queries is necessary, possibly $O(N)$ where N is the number of elements in the database (this depends on the choice of the random selector function). If fewer than $k - 1$ elements are known the database is secure under randomized queries of type average.

BIBLIOGRAPHIC NOTE

Most of the material in this section is from Leiss.[51,52,57]

2.3.2.2. *Medians, Maxima, and Minima*

Let us now see whether we can apply the method of randomizing, which we discussed in the last section for averages, in the case of selector functions, too. For simplicity we will restrict our attention to three different selector functions, namely, medians, maxima, and minima. Other suitable selector functions follow easily.

We first observe that a faithful implementation of the scheme employed for averages fails in the case of selector functions. This can be seen as follows. Let MED_k' (MAX_k', MIN_k') be the median (maximum, minimum) of $k + 1$ elements but the user supplies only the first k indices and the last element is determined by a uniform random selector function S. Let us now repeat a query of this type n times and denote the responses by $q_1, \ldots, q_n$. Suppose that our median of k elements is defined in such a way that for even k it returns the $(k/2 + 1)$th largest element (if we adopt the other possible convention, returning the $(k/2)$th largest element, the proof is analogous). Thus for sufficiently large n about half of the q_i will be the actual median of the k elements (on the average) and for the other half the returned value will be smaller if k is even and larger if k is odd. This guarantees compromise for medians on the average, for having identified the correct median we now are dealing with (straight) selector functions and hence we can apply the results of Section 2.3.1.3. For maxima and minima an even easier argument along these lines yields compromise.

Clearly the main problem is that the very restricted choice for medians, maxima, and minima allows us to filter out the contribution of the selector function S. (Note that ignoring cases where the actually returned value is the randomly chosen one, there are only two possibilities for medians, namely, the actual median of the k elements and either the next-larger (k odd) or the next-smaller (k even) element. For maxima and minima the situation is even worse; ignoring cases where the randomly chosen value is returned there is only one possible value!) Thus the logical consequence is to prohibit the possibility of filtering. This can be done by assuming that S is not a uniform random selector function but a pseudorandom selector function PS. This is a "random" selector function which depends on the indices $i_1, \ldots, i_k$ of the query in the following sense: If we repeat the query with the same indices $i_1, \ldots, i_k$ or any permutation of them, PS will return the same value. In the following we will indicate why this method yields a secure database. We will also discuss the implications of applying this modification of pseudorandomizing to averages instead of using true random selector functions. Finally, we will indicate how to obtain a simple but effective and efficient pseudorandom selector function which satisfies the necessary requirements.

First we state the main result of this section.

Theorem 19. Assume that no element of the database is known. It is not possible to compromise the database with pseudorandomized queries of type median, maximum, or minimum.

Proof. As before we have to distinguish two possible attacks, directly determining an element, and filtering. Let us first show why we can exclude attacks based on directly determining an unknown element. To determine an element of the database means that we must be able to identify some s in $\{1, \ldots, N\}$ and some response y to one of the queries such that

$$\mathrm{DK}(s) = y$$

However, since we can never be certain whether the response to a query is influenced by the pseudorandomly selected element or not, this associating of an index s with the corresponding value $\mathrm{DK}(s)$ is impossible. Now we want to verify that there is no way to remove the error by filtering techniques. First of all we observe that repeating a query does not yield any new information as the pseudorandomly selected element will be the same thus the response to the query will also be the same. Therefore removing the error from a single response is not possible. That this is also impossible for more than one query can be derived from the fact that taking medians is a rather discontinuous operation. To illustrate: Assume

$$z_1 < z_2 < \cdots < z_k$$

and let z_m be the median of these k elements where $m = k/2$ if k is even and $m = (k + 1)/2$ if k is odd. Let x be the result of the pseudorandom selector function PS. If k is even the response does not change at all as long as $x \leq z_m$ (no matter how small x is!). For $z_m < x < z_{m+1}$ the response will be x and it will be z_{m+1} for all $x \geq z_{m+1}$. For odd k the situation is reversed; as long as $x \geq z_m$ the response does not change, for $z_{m-1} < x < z_m$ it will be x and for all $x < z_{m-1}$ it will be z_{m-1}. Similar arguments hold for maxima and minima. This kind of behavior together with the fact that we are dealing with pseudorandom selector functions yields the claim that filtering techniques for removing the "noise" introduced by randomizing are very unlikely to yield compromise. □

The reader should be aware that this argument is not entirely tight; proofs of security are notoriously difficult (unless they are proofs by contradiction) for they require anticipating yet unknown situations and attacks. We do feel confident, however, that the proposed method will stand the test of time.

It should be pointed out that the method is extremely sensitive to modifications. As an example consider the seemingly innocent additional assumption that the value returned by a query be that element of the original k values which is closest to the pseudorandomized answer. In this case the database can be compromised because the queries are now of the type arbitrary selector functions where Theorem 12 guarantees compromise in at most $k^2 + 1$ queries.

One problem with the above theorem is the assumption $D_0 = \emptyset$. While it can be safely expected that knowing a few data elements does not enable the user to compromise the database it appears possible that the database can be compromised if $O(k)$ elements are known. However, no obvious method is known to achieve this compromise unconditionally and in all cases (but see the exercises).

Let us remark on this method of using pseudorandom selector

functions in contrast to randomized queries of type average. Basically from the point of view of security the pseudorandom approach is better. This can be seen from the fact that the problem in the first part of Theorem 18 does not exist any longer; knowing $k - 1$ or more data elements would thus not guarantee compromise as the averaging described there for this case would not be possible any more. On the other hand we feel that for the sake of accuracy of the responses the (truly) randomizing method is to be preferred. This and related topics will be discussed in the next section.

Finally, we indicate how to implement a pseudorandom selector function which returns the same value whenever the same sequence of indices or a permutation is presented. Let $(i_1, \ldots, i_k)$ be the query and let PS be the pseudorandom selector function. Then an easy way to satisfy the condition is by taking $\mathrm{DK}(i_1) + \cdots + \mathrm{DK}(i_k)$ as seed to the random selector function S mentioned at the end of Section 2.3.2.3. This is a very inexpensive way to obtain a pseudorandom selector function and can be implemented in virtually any presently used database yet it satisfies all the necessary requirements.

Questions and Exercises

1. Assume the $k - 1$ largest and the $k - 1$ smallest elements of the database are known. Show that in this situation global selective compromise is guaranteed provided the database elements are pairwise distinct. (Recall that we are using pseudorandomized queries!)

2. Give an example which demonstrates that the assumption in 1. that the database elements be pairwise distinct cannot be dropped if compromise is to be guaranteed.

3. Assume that the g largest and the h smallest elements of the database are known. Furthermore, assume that the database elements are not (known to be) pairwise distinct. Show that on the average global selective compromise can be achieved with a degree of confidence depending on g and h [more specifically directly depending on $g - (k - 1)$ and $h - (k - 1)$].

Bibliographic Note

Some of the material presented in this section can be found in Leiss.[52]

2.3.2.3. Restricted Randomizing

In the last two sections (2.3.2.1 and 2.3.2.2) we introduced the method of randomizing for queries of the types averages and certain selector functions. Applying this method to statistical databases with these queries results in databases which can be shown secure under the queries involved. Thus the main objective, namely, secure databases, has been achieved. However, at no point in the exposition did we address the question at what price this security is attained. This will be done in the present section. First we discuss the accuracy of the resulting responses; only empirical results are of any use since theoretically the accuracy can be arbitrarily bad. We then discuss a method which increases the accuracy of the answers but at a cost in the security of the resulting database. In effect we will be discussing a security–accuracy trade-off; we feel that this makes the method very attractive as not all data are equally confidential. This is enhanced by the fact that the trade-off can easily be parametrized; varying a parameter allows to increase or decrease the accuracy and at the same time decrease or increase the security.

Let us first direct our attention to the question of accuracy from a purely theoretical point of view. We claim the following:

Proposition 7. The accuracy of randomizing can be arbitrarily bad.

Proof. We will show this for randomized queries of type average and of type median. First averages. Let y be the (true) average of k data items and let y' be the randomized response, i.e., y' is the average of $k + 1$ elements consisting of the original k elements plus one new element $\mathrm{DK}(s)$, namely, the one selected by the random selector function S. Let M be an arbitrary positive number. We claim that the relative as well as the absolute error can be greater than M. Assume $y = 1$ and let $\mathrm{DK}(s) \geq (M + 1)(k + 1)$. Clearly the relative error is greater than M, as is the absolute error. Since this holds for all $M > 0$ the claim follows for queries of type average. We proceed with the proof for queries of type median. Again let M be any positive number, let y be the true median of k elements, and let y' be the randomized response. We show that the relative and the absolute errors can be greater than M. Assume the k elements are as follows [we assume the median is the $(k/2 + 1)$ largest element]: Let k be even and let the $k/2$ smallest elements be 1 and the rest be $M + 2$. In this case the (true) median is 1. If the randomly chosen value is at least $M + 2$ the value of y' is $M + 2$. Thus absolute as well as relative errors are greater than M. □

The analogous results for maxima and minima are even easier to obtain.

While these results are certainly correct, from a practical point of view they are meaningless and therefore best ignored. Of far greater practical significance are the results of a simulation. Such a simulation was done for averages and medians. Queries are simulated by k calls to a uniform random generator which returns a value within a certain interval (here [0, 500]). This is done for various values of k (namely, k — 5, 10, 20, 50, 100) and is repeated 10,000 times for each k. The results are summarized below in Table 2.2. For more details the reader is referred to Leiss.[52]

While for most values of k the average relative errors are certainly acceptable, this is not the case for the maximal errors, primarily for smaller values of k and especially for queries of type median. Consequently it would be a big improvement of the method in general if one could reduce the maximal relative error in one way or another. The method of restricted randomizing does just that.

Recall that in the original method of randomizing the results of S and PS are arbitrary elements of the database, i.e., DK(s) for some s in $\{1, \ldots, N\}$. Thus DK(s) can differ by an arbitrary amount from the precise response to the query (i.e., the query without randomizing), which in turn can affect greatly the response to the randomized query.

Table 2.2. Accuracy with Randomizing

	Averages			Medians	
k	Average relative error (%)	Maximal relative error (%)	k	Average relative error (%)	Maximal relative error (%)
5	8.9	138.0	5	14.6	1380
10	4.7	26.0	10	8.4	82
20	2.4	11.2	20	4.5	55
50	1.08	3.2	50	2.0	31
100	0.5	1.4	100	1.0	16

This suggests to restrict the range out of which the results of S and PS are to be selected in such a way that they do not differ too much from the precise response. It is not difficult to see that this can create problems as far as security is concerned; if one restricts this range too much it is quite conceivable that the randomized response is always the precise response, which in turn implies that compromise is now possible. On the other hand it must be recognized that not all data are equally secret; consequently a method would be desirable where the owner of the data is able to set a parameter which determines the accuracy of the responses and at the same time the security of the database. The method described below satisfied all these requirements. It can be employed for random selector functions as well as for pseudorandom selector functions. It is easy to implement and to use. Thus we feel that it is very attractive for anybody who wants to design a statistical database.

Consider a query $(i_1, \ldots, i_k)$ of type f where f is either average or one of the selector functions median, maximum, minimum; let q be its (true) response:

$$q = f(\mathrm{DK}(i_1), \ldots, \mathrm{DK}(i_k))$$

In our randomized model the response will not be q but q' defined by

$$q' = f(\mathrm{DK}(i_1), \ldots, \mathrm{DK}(i_k); \mathrm{DK}(s))$$

where $\mathrm{DK}(s)$ is the result of S or PS (s in $\{1, \ldots, N\}$). As pointed out q' can differ arbitrarily from q. Let mx (mn) be the largest (smallest) of the $\mathrm{DK}(i_j)$, $j = 1, \ldots, k$. Instead of allowing $\mathrm{DK}(s)$ to be arbitrary we assume that $\mathrm{DK}(s)$ is required to satisfy the following inequalities:

$$q - (\mathrm{mx} + \mathrm{mn})/(2j) \leq \mathrm{DK}(s) \leq q + (\mathrm{mx} + \mathrm{mn})/(2j)$$

for some $j > 0$ suitably chosen. This method of selecting the randomly chosen element will be called restricted randomizing. Clearly the crucial point is the choice of j. If j is too small ($j \ll 1$) then for all practical purposes we will end up with unrestricted randomizing; if j is too large the contribution of $\mathrm{DK}(s)$ will not change q at all thereby rendering the method useless as now the database can be compromised. Furthermore if j is too large it is possible that no $\mathrm{DK}(s)$ satisfies the required inequalities. This can happen for f being average; for f being one of the selector functions the worst which can occur is that $\mathrm{DK}(s)$ must be exactly q. In these and similar cases, alternative schemes must be provided. The following method was found to be very effective. Let S be as usual the uniform random selector function described in Section 2.3.2.1. Given a query $(i_1, \ldots, i_k)$ of one of the four types average, median, maximum, minimum, we call the function S; let x be the result of this call. (Clearly if the type f is one of the selector functions we will call S with the appropriate seed as described in Section 2.3.2.2.) Then we test whether x satisfies the inequalities; if yes then we use x in the computation of q' otherwise we continue calling S (regardless of the original seed!) until either the result does satisfy the inequalities or a certain preset number of calls has been made (e.g., $20j$) in which case the value to be used in the computation of q' is to be the one which came closest to satisfying the inequalities. The reader should note that with this modification we will

Table 2.3. Accuracy with Restricted Randomizing

		Averages		Medians	
k	j	Average relative error (%)	Maximal relative error (%)	Average relative error (%)	Maximal relative error (%)
5	1	8.2	65	13.8	1505
	2	5.8	54	11.2	1281
	5	2.8	28.5	6.5	294
	10	2.1	21.3	3.6	215
	20	1.9	23.4	2.0	84
	50	1.8	17.1	0.8	38
10	1	4.7	25.9	8.5	81
	2	3.9	19.4	8.2	83
	5	1.8	14.0	5.9	74
	10	1.2	12.0	3.7	56
	20	1.04	12.9	2.2	64
	50	0.99	10.4	1.0	17
20	1	2.4	11.2	4.5	56
	2	2.2	10.9	4.5	59
	5	0.93	4.6	3.9	49
	10	0.6	3.9	3.0	42
	20	0.47	3.8	1.9	20.5
	50	0.43	3.4	0.9	11.4

still end up with a pseudorandom selector function if we started with such a function.

In order to test the accuracy of the results another simulation was done exactly like the first one but implementing the above-described method of restricted randomizing. The results for $k = 5, 10, 20$ are given above in Table 2.3 for various values of j, namely, $j = 1, 2, 5, 10, 20, 50$.

Finally, we conducted an experiment to establish that these methods indeed result in a secure database. Again this was done by way of a simulation. This simulation is compatible with all the others mentioned here. In order to determine how safe the resulting database is we determined the probability that the computed value of any (compromised) database element is afflicted with an error of a certain magnitude. Method and results of this very extensive simulation (several hundred hours of CPU time on a Honeywell 66/60) are discussed at length in Leiss[57]; here we present only a small part of the tables obtained. In particular, we will assume that all queries involved in the compromise contain 20 indices, i.e., $k = 20$. Then Tables 2.4 and 2.5 give the probability that the computed value of any database element will have a certain error, Table 4 if the queries involved in the compromise are posed just once (no filtering) and Table 5 if the queries are repeated 100 times. More spe-

Table 2.4. Probability of Errors of Computed Values, $k = 20$, No Repetition of Queries

Interval of error %	j					
	1	2	5	10	20	50
[0,1 [	1.14	1.20	3.14	5.93	12.54	21.76
[1,2 [	1.16	1.23	2.90	6.27	12.19	20.05
[2,4 [	2.21	2.50	6.40	12.97	21.13	30.03
[4,8 [	4.36	4.80	12.36	23.18	27.98	9.29
[8,16 [	8.97	9.81	23.77	26.56	11.90	7.49
[16,32 [	18.37	19.55	25.75	12.39	6.32	5.14
[32,64 [	28.35	27.96	12.75	6.17	3.85	3.03
[64,128[	17.71	16.39	6.27	3.26	1.92	1.35
[128, – [	17.73	11.13	6.64	3.28	2.17	1.87

Table 2.5. Probability of Errors of Computed Values, $k = 20$, 100 Repetitions of Queries

Interval of error %	j					
	1	2	5	10	20	50
[0,1 [	0.15	0.10	1.07	7.70	20.02	22.65
[1,2 [	0.12	0.14	1.69	7.79	13.57	24.78
[2,4 [	0.53	0.55	6.38	16.94	34.08	25.19
[4,8 [	2.39	2.90	24.00	39.12	13.63	8.27
[8,16 [	14.57	18.37	35.40	13.56	7.18	7.52
[16,32 [	37.09	36.69	15.33	6.93	4.96	5.05
[32,64 [	22.68	20.44	7.73	3.79	3.16	3.25
[64,128[	10.94	10.18	4.12	2.02	1.54	1.52
[128, – [	11.54	10.66	4.25	2.12	1.69	1.78

cifically, if we are using queries with $k = 20$ and we attempt to compromise, i.e., compute the value of a particular database element, the expected or probable error which this computed value will have (owing to the fact that the queries are randomized) is given in Tables 4 and 5. For example, if we know that the randomizing factor j is 10, then the probability that the computed value for the element is within 1% of the real value of the database element is 5.93%, the probability that the computed value has an error of 1%–2% is 6.27%, etc. If we want to know the probability with which the computed value lies within 4% of the real value we must add up 5.93%, 6.27%, and 12.97% to yield 25.17%.

The results suggest that restricted randomizing for a suitable choice of j is a very useful way of protecting confidential data on the one hand while on the other hand providing meaninfgul statistical information based on these confidential data elements.

Bibliographic Note

The method of restricted randomizing was introduced in Leiss.[52] This paper and Leiss[57] also contain a detailed description of the simulations cited here and a more extensive discussion of the results.

2.4. Conclusion

Most of Section 2, namely, Sections 2.2.1, 2.2.2, and 2.3.1, is devoted to showing that security is a very elusive property of a statistical database and that the results moreover are often counterintuitive. In particular, purely syntactic restrictions on the (form of the) queries (like overlap) either do not work at all (e.g., for nonzero overlap) or are entirely unenforceable (e.g., for zero overlap). This negative conclusion also holds for most other schemes (see, for instance, Denning and Denning[23] for a good introduction) which are suggested or employed in order to obtain secure statistical databases. We therefore believe that the method described in the last section (2.3.2) is of considerable practical importance, as to the author's knowledge it is the only method which satisfies the following key requirements:

(a) It guarantees the security of the underlying database.

(b) It can be implemented for a variety of commonly used query functions.

(c) The implementation can be added to existing statistical databases without difficulties.

(d) The cost per query is completely negligible.

(e) The method does not make any unreasonable assumptions; in particular it works if the user knows a (relatively small) number of data elements.

(f) It can also be used for small databases. These databases typically are the most insecure ones; the larger a database, the more amenable it is to applications of traditional methods such as those employed by the census bureau (sampling techniques).

The method if implemented as suggested furthermore allows to decrease the security of a statistical database (in a continuous fashion) which results in a gain in the accuracy of the responses. Its main disadvantage is the fact that the answers are not precise (although it is debatable whether it makes sense from a practical point of view to talk at all about *precise* answers in a statistical database); instead the responses of the queries are slightly modified. However, the discrepancy between the precise answers and the actual responses appears quite acceptable for virtually all applications. Therefore we feel that this method is superior for all practical purposes to any of the schemes suggested elsewhere.

Chapter 3

Authorization Mechanisms

3.1. Introduction[†]

The concept of authorization is central to any act of data manipulation which retains at least some vestiges of privacy or security of information. Very briefly, the basic question is "who may do what with a certain data object?" For our purposes, however, this is much too general; for example, it would include all the problems addressed in Chapter 2. In this chapter we will concern ourselves with the following situation. Any user of a data object has certain rights to it. We will discuss how such rights may be acquired, how they can be passed on to other users, whether they can be revoked, and how (if at all possible) a user can determine whether another user has the possibility in a given situation to exercise certain rights to this object. Therefore this chapter is concerned with the granting, the revoking, and the administration of certain privileges.

Before we more precisely outline the exact limits of our investigation let us first address an implicit assumption in this scenario which ought to be spelled out quite clearly. This is the underlying supposition that users are what they appear to be. In other words, we assume that we have a correctly working user identification procedure and we assume that each user can in fact be unmistakably identified. This assumption itself has a number of important implications. For one, the software

[†] This introduction is of a more theoretical nature. As we will discuss a practical authorization system in Section 3.4 we will postpone the motivation from this point of view until that section.

which performs this identification works correctly and cannot be accessed in any other way by other users. For another, only the real user A can identify him/herself as user A. Thus an attempt by any user other than A at claiming to be A will be detected. For example, if the identification is performed via a password, only user A is assumed to know his/her password and nobody else. This and related issues will be addressed in Chapter 4 (Cryptosystems).

Another point which we will not address further in the following is the problem of correctness and security of the underlying software system, in particular, the operating system. One reason for excluding this topic is that this issue is usually outside of the users' influence. Another reason for this exclusion is the fact that a solution to this problem could fill another book. Intuitively it should be clear what is meant by correctness: The system performs exactly as specified in its formal description. While this is rarely satisfied in practice (all commercially used operating systems have a multitude of inconsistencies and errors; they "offset" this, however, by a lack of formal description in the first place), the reader should be aware that even this does not imply that the system is secure. It may well be that in the formal description some arcane way of obtaining access had not been excluded and is thus allowed, access which intuitively should not have been permitted and which in fact was to be disallowed in the system according to the intentions of the designers. Clearly these difficulties will increase with the size and the complexity of the software system in question. It is therefore interesting to note that there are provably secure operating systems (see Neumann,[67] Millen,[64] Popek and Kline,[69–71] Popek and Farber,[72] and Schroeder *et al.*[85]). While it must be said that these operating systems are still very primitive and furthermore rather inefficient, it is important for us to note that it is indeed possible to write such systems. Consequently it does make sense to assume such a secure operating system for the considerations of this part. It is also in the spirit of top-down design of large systems to assume the correctness of individual components when one attempts to demonstrate the correctness of the whole system. For a concise discussion of the issues involved in the design of secure operating systems and security kernels see Jones.[44]

Another problem which will be excluded from the discussion for obvious reasons is the physical protection of hardware, software, and information involved. For a thorough review of the overall principles of computer security which also includes this issue see Gaines and Shapiro.[33]

Our main objective of this part then is to determine whether a user is authorized to access a data object in a certain prespecified way, and if not how to prevent this unauthorized access within the framework of a given system (model).

Before we proceed we want to clarify what kind of users and what kind of data objects we have in mind. In Chapter 2, Statistical Database Security, the user was an actual person (or possibly a program) and the data object was a file ordered and accessible in a certain way, with the essential information being numbers. In this part we will generalize both user and data object. A user will now be any entity which acts on data objects, while a data object may be anything stored in some computer accessible medium. In particular, both user and data object may be programs. Thus the names "user" and "data object" only imply the direction of an action. It is quite conceivable that at some time later the present user will become the data object and vice versa; this situation may be illustrated by the actions of coroutines A and B where first A modifies B and then passes control to B, which itself proceeds to modify A in turn.

In view of this the most important notion is not any longer access but the passing of control. Thus what is to be analyzed when discussing the security of an authorization system (as we will call these systems) is the flow of control. That this is by no means trivial is illustrated by the following sequence of statements:

```
if x ≠ 0 then go to L;
     print (0);
  L: print (1);
```

While it is quite obvious that this program will communicate whether $x = 0$ or $x \neq 0$, it is by no means easy to formulate general and at the same time precise rules according to which this program is detected as a potential security risk if the value of the variable x is assumed to be kept secret to the person who is to read the output. (For a partial solution to this problem see Reitman and Andrews;[78] also Reitman[77]). Most implemented solutions employ the usual military policy for classifying information. It can be stated as follows: There is a two-dimensional hierarchy of security classes, one component (i) being the authority level, the other one (a) being the category. Typical authority levels are confidential ($i = 1$), secret ($i = 2$), and top secret ($i = 3$). A category is one or more compartments, typical compartments being unrestricted ($a = u$), restricted ($a = r$), sensitive ($a = s$), and crypto ($a = c$). Information can be passed from a user A with security class (i, a) to a user B with security class (j, b) iff $i \leq j$ (i.e., B's authority level is not lower than A's) and a is contained in b (i.e., every compartment associated with A is also associated with B). Several systems have been written based on this scheme or some variant of it and are actually operational (see, for instance, Weissman,[97] Millen,[65] Walter,[96] and Neumann[67]).

The Bell–La Padula *-property is an interesting variant; it requires that if a subject has read permission to an object a and write access to an object b at the same time, the security class of a must be dominated by that of b (see Bell and La Padula[3]).

There are some serious problems with any of these approaches. The most serious one is the fact that information will always move upward and never downward. While this is certainly desirable in most cases it also has the result that after some time much information will be overclassified. While there are ways to alleviate this problem, they either are not quite consistent with the objective of security or they enjoy only very limited success in preventing overclassification. Another very serious problem is the lack of generality. Consider, for example, a simple sorting routine S. It turns out that for each security class one needs a separate copy of S. For let S's security class be (i, a) and let (j, b) be the security class of the file F to be sorted. If S wants to read F then $j \leq i$ and b must be contained in a. If S wants to write F then $i \leq j$ and a must be contained in b. This leads to an extreme redundancy. Also notice that problems are likely to increase with the size of a system.

Despite the practical difficulties, we already pointed out that at least in some instances these problems have been effectively solved. Our interest will concentrate on the "safety problem" which essentially is embodied in the question "Can user A ever get access to data object B?" Naturally, if at any given time only one user has all rights to a given data object this question will always be trivially solvable. The problems start, however, as soon as we are dealing with shared data objects. In the following we will highlight two approaches which represent the two extremes of the spectrum. The first one, due to Harrison, Ruzzo, and Ullman, shows that the operations of a realistic access control scheme are sufficiently powerful to render the safety problem undecidable. On the other hand, the second approach, commonly referred to as "grant–take" method, has not only a decidable safety problem but in fact one which can be solved in linear time.

We conclude this chapter on authorization mechanisms by giving a brief account of a commercially implemented system which shows that the abstract theory is by no means very far removed from reality although it does not yet completely model all real problems. Thus we submit that we have here a case at hand where a theoretical development does have some influence on practical implementations.

Let us terminate this introduction with a short and informal description of an authorization system; the formal definition will follow later. There is a finite number of objects of different types. The types are fixed beforehand, for instance, subjects or data objects, but the number of objects is usually not fixed, at least not in the systems where it is possible to create new objects (most useful systems have this property). The objects are related by rights which some object may have to some other object. The rights may be read, write, delete, execute, update, etc. Thus, object X may have the rights to read and to run the object Y which in turn may have the right to read X and the right to write Z. A question central to these authorization systems is whether giving object X some right r over another object Y will enable other users to gain a certain right r' to some object Y'. This problem is commonly referred to as the safety problem for authorization systems. It will be the main issue of the following two sections.

3.2. The Undecidability of the Safety Problem for General Authorization Systems

In this section we will show that there is no systematic method to determine in a given situation within a given authorization system whether a user, henceforth called a subject, can obtain a particular right to a certain data object; thus we will show that the safety problem is undecidable. This result is due to Harrison, Ruzzo, and Ullman; it clearly limits the theoretical possibilities for answering the question posed in the general introduction to this book, which is the motto of this and the subsequent chapter, namely, "If I give A this right and B that right will that enable C to do something terrible to me?" Before we present the underlying model and formulate the result we will discuss what it means, on the one hand for the theorist who would like to attain a uniform theory of security together with universally applicable methods, on the other hand for the practitioner who is concerned with the safety problem for *his/her* particular system, maybe even in a very specialized situation. The theorem itself will be obtained by showing that the authorization system can simulate an arbitrary Turing machine in such a way that an answer to the safety problem is equivalent to this (simulated) Turing machine, started in the initial configuration, eventually entering a certain state. Since this problem for arbitrary Turing machines is known to be undecidable so is the safety problem for (general) authorization systems.

3.2.1. Relevance and Implications of the Result

The undecidability of the safety problem is quite surprising. It also is somewhat discouraging; the fact that a problem P is undecidable tends to lure people into believing that P cannot be solved. This is incorrect. Problem P is undecidable only means that there is no single algorithm which will work in all instances of P. It does not mean that given a certain instance of P we cannot solve P for this case, nor does it mean that there cannot be large classes for which P can be solved. Usually all this is possible. As an illustration let us consider the problem whether two given context-free grammars generate the same language. It should be clear that this is a problem of considerable practical importance as it is related to the parsing and compiling of programs. Nevertheless it is known that this problem is undecidable, i.e., it is impossible to find a definite method which will answer this question for all possible pairs of context-free grammars within a finite number of steps. Note that given two grammars G_1 and G_2, generating the languages L_1 and L_2, respectively, there are only two possibilities, either

$$L_1 = L_2 \qquad \text{or} \qquad L_1 \neq L_2$$

Furthermore if $L_1 \neq L_2$ then we can in fact exhibit a word for which we can show that it is in one of the two languages but not in the other, and we can do this with a finite number of steps. That this problem is undecidable only says that we have no general method for all such pairs (G_1, G_2). However, in virtually all given specific instances it will be possible to determine whether $L_1 = L_2$ or not (although this may require a good deal of ingenuity on the part of the solver).

The situation for the problem at hand is completely analogous. No claim is made about our ability to answer a specific instance of the safety problem. All that is implied by its undecidability is the fact that no single algorithm will suffice to answer it for all possible instances. In practical terms then this result does not imply that any attempt at solving the problem for certain instances is futile. On the contrary, the result suggests two avenues: On the one hand one might try to devise an algorithm to answer the safety problem for a very specific kind of authorization system, typically one which has already been implemented and is being used and studied. On the other hand one might try to identify classes of authorization systems for which one is able to prove that they have a decidable safety problem. Both cases can again be illustrated by the equivalence problem for context-free languages. The first approach would correspond to a situation where a compiler designer, who needs to implement a particular programming language, modifies its grammar in order to facilitate parsing and then has to prove that the two grammars generate the same language. The second approach corresponds to making certain general assumptions about the two context-free languages; one subclass for which the equivalence problem is known to be decidable is the one defined by the restriction that one of the two context-free languages be regular. Finally, it is unknown (as of Summer 1981) whether the equivalence problem is decidable if the two context-free languages are assumed to be deterministic; this would relate to certain authorization systems for which we do not know whether they have a decidable safety problem.

3.2.2. The Model and Its Safety Problem

For the purposes of this section an authorization system will consist of a finite set R of *generic rights* and a finite set C of *commands*. Any command c in C is of the following form:

command $c(X_1, \ldots, X_k)$
    if r_1 in (X_{s_1}, X_{o_1}) **and**
        ⋮
        r_m in (X_{s_m}, X_{o_m})
    then
        op_1
        ⋮
        op_n
    end

for $m \geq 0$ (for $m = 0$ we simply have an unqualified list of operations $\text{op}_1, \ldots, \text{op}_n$).

Each op_i is one of the following six primitive operations:

$$\begin{array}{l} \text{enter } r \text{ into } (X_s, X_o) \\ \text{delete } r \text{ from } (X_s, X_o) \\ \text{create subject } X_s \\ \text{destroy subject } X_s \\ \text{create object } X_o \\ \text{destroy object } X_o \end{array}$$

where $X_1, \ldots, X_k$ are formal parameters, $r, r_1, \ldots, r_m$ are generic rights in R, and $1 \leq s, s_1, \ldots, s_m, o, o_1, \ldots, o_m \leq k$.

Given an authorization system (R, C) we can define an instantaneous description of (R, C) called a *configuration*. A configuration of (R, C) is a triple (S, O, P), where O is the set of current objects, S is a subset of O and denotes the set of current subjects, and P is an access matrix with a row for every s in S and a column for every o in O. The entry $P[s, o]$ is the set of rights in R which the subject s has to the object o; consequently $P[s, o]$ is a subset of R. Note that the row $P[s, *]$ indicates what s can do with all the objects in O (capability list) and the column $P[*, o]$ indicates what subjects have rights to the object o and what these rights are (access list).

Intuitively the six primitive operations have precisely the meaning suggested by their names. However, as we describe the status of our system in terms of access matrices it is necessary to define formally their effect on these matrices. Let (S, O, P) and (S', O', P') be configurations of the authorization system (R, C), and let op be a primitive operation. Then

$$(S, O, P) \quad \Rightarrow_{\text{op}} \quad (S', O', P')$$

if

(1) op is enter r into (s, o) and
$S = S'$, $O = O'$, s in S, o in O,
$P'[s', o'] = P[s', o']$ for $(s', o') \neq (s, o)$, and
$P'[s, o] = P[s, o] \cup \{r\}$

(2) op is delete r from (s, o) and
$S = S'$, $O = O'$, s in S, o in O,
$P'[s', o'] = P[s', o']$ for $(s', o') \neq (s, o)$, and
$P'[s, o] = P[s, o] - \{r\}$.

(3) op is create subject s'
with s' not in O (i.e., s' is not a current object name)
$S' = S \cup \{s'\}$, $O' = O \cup \{s'\}$,
$P'[s, o] = P[s, o]$ for all (s, o) in $S \times O$,
$P'[s', o] = \emptyset$ for all o in O', and
$P[s, s'] = \emptyset$ for all s in S'.

(4) op is create object o'
with o' not in O (o' not a current object name),
$S' = S$, $O' = O \cup \{o'\}$,
$P'[s, o] = P[s, o]$ for all (s, o) in $S \times O$, and
$P'[s, o] = \emptyset$ for all s in S.

(5) op is destroy subject s' and
s' in S, $S' = S - \{s'\}$, $O' = O - \{s'\}$, and
$P'[s, o] = P[s, o]$ for all (s, o) in $S' \times O'$.

(6) op is destroy object o' and
o' in $O - S$, $S' = S$, $O' = O - \{o'\}$, and
$P'[s, o] = P[s, o]$ for all (s, o) in $S' \times O'$.

Now we can describe how a command is executed in an authorization system (R, C). Let Q be a configuration. We say that Q derives Q' under the command c and the actual parameters $x_1, \ldots, x_k$, with c given by

command $c(X_1, \ldots, X_k)$
 if r_1 in (X_{s_1}, X_{o_1}) **and** ... **and** r_m in (X_{s_m}, X_{o_m})
 then $\text{op}_1, \ldots, \text{op}_n$ **end**

and we write

$$Q \vdash c(x_1, \ldots, x_k)\ Q'$$

if Q' is defined as follows:

(1) If the condition of the command c is not satisfied (i.e., there is some i in $\{1, \ldots, m\}$ such that r_i is not in $P[x_{s_i}, x_{o_i}]$), then $Q = Q'$.

(2) If the condition of the command c is satisfied then there must exist configurations $Q_0, \ldots, Q_n$ such that

$$Q = Q_0 \ \Rightarrow \text{op}_1^* \ Q_1 \ \Rightarrow \text{op}_2^* \ \cdots \ \Rightarrow \text{op}_n^* \ Q_n$$

where op_i^* denotes the primitive operation op_i with the actual parameters $x_1, \ldots, x_k$ substituted for the formal parameters $X_1, \ldots, X_k$, respectively. Then $Q' = Q_n$.

Furthermore we write

$$Q \vdash_c Q'$$

if there exist (actual) parameters $x_1, \ldots, x_k$ such that Q derives Q' under c and these actual parameters, and we write

$$Q \vdash Q'$$

if there exists a command c such $Q \vdash_c Q'$. Finally, if we can obtain Q' from Q by zero or more repetitions of $\vdash$ we write

$$Q \vdash^* Q'$$

formally $\vdash^*$ is the reflexive and transitive closure of the relation $\vdash$.

We now come to the safety problem. Informally it can be stated as follows. Suppose we are in some configuration Q_0. We wish to determine whether there is a sequence of commands $c_1, \ldots, c_n$ which, when applied to Q_0, yields another configuration Q such that execution of one more command in configuration Q will result in entering a right r into a cell of the access matrix which previously did not contain this right r. More formally we say a command $c(X_1, \ldots, X_k)$ *releases* the generic right r from configuration $Q = (S, O, P)$ if there are some actual parameters $x_1, \ldots, x_k$ such that (1) the condition of c is satisfied in Q, and (2) if the primitive operations of c are $\mathrm{op}_1, \ldots, \mathrm{op}_n$ then there exists an m in $\{1, \ldots, n\}$ and there exist configurations

$$Q = Q_0, Q_1, \ldots, Q_{m-1} = (S', O', P') \qquad \text{and} \qquad Q_m = (S'', O'', P'')$$

such that

$$Q_0 \Rightarrow_{\mathrm{op}_1^*} Q_1 \Rightarrow_{\mathrm{op}_2^*} \cdots \Rightarrow_{\mathrm{op}_m^*} Q_m$$

and

$$r \text{ not in } P'[s, o] \text{ but}$$
$$r \text{ in } P''[s, o] \text{ for some } s \text{ in } S' \text{ and } o \text{ in } O'$$

(op_i^* denotes op_i after substitution of the actual parameters.)

Clearly it can always be assumed that op_m is enter r into (s, o). Given Q, c, and r, it is easily determined whether r is released or not. We remark that in the original paper "leak" is used instead of our "release." However, as leak has a negative connotation and the release of a right is not necessarily bad, in fact frequently desired, we opt for a neutral word. Note that anytime we give a right to somebody we release this right; any interesting system will have a command which will allow this kind of sharing.

Now the formal definition of safety is easy. Given an authorization system (R, C) and a generic right r in R we say that the (initial) configuration Q_0 is unsafe for r if there is a configuration Q such that $Q_0 \vdash^* Q$ and a command c such that c releases r from Q. If Q_0 is not unsafe for r we call Q_0 safe for r.

Example. Suppose we have an authorization system (R, C) which at some point in time has eight objects and four subjects. The subjects are OS (operating system), DBM (database manager), COB1 and COB2 (two COBOL programs), and the remaining objects are SAL80, EMP79, EMP80, INVENT, and OUTPUT (files with the salaries for 1980, the employment for 1979 and 1980, the inventory and the output, respectively). Let the following generic rights be in R:

read (r),
write (w),
execute (e),
update (u), and
delete (d)

The operating system (by definition) has all rights to all the objects including itself and the database manager has the right e to OS, the rights r, w, u, and d to the two employee files, the salary file, the inventory file, and the output file. Finally, the program COB1 has the rights r, w, u, and d on SAL80, INVENT, and OUTPUT, and COB2 has the rights r, w, u, and d on EMP80, but only the right r on EMP79. The corresponding access matrix P would be as follows:

	OS	DBM	COB1	COB2	SAL80	EMP79	EMP80	INVENT	OUTPUT
S	all	all	all	all	all	all	all	all	all
BM	*e*	—	*er*	*er*	*rwud*	*rwud*	*rwud*	*rwud*	*rwud*
OB1	—	—	—	—	*rwud*	—	—	*rwud*	*rwud*
OB2	—	—	—	—	—	*r*	*rwud*	—	—

Now suppose one of the commands in C is

$c(X_1, X_2, X_3, X_4)$
if e in $P[X_1, X_2]$
  and e in $P[X_2, X_3]$
  and r in $P[X_3, X_4]$
then
  enter e into $P[X_3, X_1]$
end

We want to know whether COB2 can ever delete COB1. In this case the safety problem can be answered; it is indeed possible for COB2 to delete COB1 in the given situation. To see this consider the following correspondence between formal and actual parameters:

X_1 OS
X_2 DBM
X_3 COB2
X_4 EMP79

It follows that COB2 (X_3) acquires the right e to OS (X_1) after execution of command c with these parameters. As OS has all rights to all objects, COB2 is in a position to cause OS to delete COB1.

3.2.3. The Main Theorem

In this section we show that the safety problem is undecidable for general authorization systems. In the proof of this theorem we will show that the decidability of the safety problem would imply the decidability of a certain problem concerning Turing machines. This problem, however, is well known to be undecidable thereby rendering the safety problem also undecidable.

For the sake of completeness we review now the notion of a Turing machine; for more details see, for instance, Hopcroft and Ullman.[42] A Turing machine TM is a quintuple

$$\mathrm{TM} = (K, T, d, k_0, F)$$

where K is the finite set of states, T is the (finite) alphabet of tape symbols ($T \cap K = \emptyset$), k_0 in K is the initial state, and the subset F of K is the set of final states. Finally, d is the move function of TM,

$$d\colon K \times T \to K \times T \times \{\text{LEFT, RIGHT}\}$$

The interpretation is as follows. The Turing machine has a tape consisting of squares numbered 1, 2, 3, . . . (i.e., the tape is unbounded to the right) and a tape head which always scans a square of the tape. The Turing machine is always in a state, initially in the state k_0. In the beginning the input to the Turing machine, a word over T but not containing the special character ƀ (blank), which is always assumed to be in T, will be on the initial squares of the tape, while all remaining squares contain ƀ. If the machine is in state k and the square scanned by its head is a, the move of TM is described by its move function d. If $d(k, a) = (k', a', \text{RIGHT})$ then TM changes from state k to state k', replaces the symbol a in the square it currently scans to a', and moves its head to the square immediately to the right. Similarly, if $d(k, a) = (k', a', \text{LEFT})$. Note, however, that TM can never move to the left of square one. (If an attempt to do this is ever made, TM will halt and reject its input.) Finally, TM accepts its input if TM eventually ends up in a final state when presented with a word over $T - \{ƀ\}$ on the initial portion of its tape.

Example. Let us construct a Turing machine M which accepts (by final state) the language

$$\{a^n b^n c^n d^n \mid n \geq 1\}$$

Informally, when presented with an input $\$w$ (with \$ a marker indicating the beginning of the word w over $\{a, b, c, d\}$), M will do the following:

(a) M checks in a left-to-right pass that w starts with a's, then follow b's, then c's, and finally w ends in d's.

(b) M then replaces (in a right-to-left pass) the last d by D, the last c by C, the last b by B, and the last a by A (last = rightmost). Then M goes to the right again until it finds the first D (first = leftmost); M then repeats this step (b) until it cannot find another d, c, b, or a. If this is the case simultaneously M accepts, otherwise M rejects its input w.

Formally we have

$$M = (\{0, 1, 2, 3, 4, 5, 6, 7, 8, 9, 10, 11\}, \{a, b, c, d, A, B, C, D, \$, ƀ\}, d, 0, \{11\})$$

with the next-move function d given by

	a	*b*	*c*	*d*	*A*	*B*	*C*	*D*	$	b̸
0	(0,*a*,R)	(1,*b*,R)	—	—	—	—	—	—	(0,$,R)	—
1	—	(1,*b*,R)	(2,*c*,R)	—	—	—	—	—	—	—
2	—	—	(2,*c*,R)	(3,*d*,R)	—	—	—	—	—	—
3	—	—	—	(3,*d*,R)	—	—	—	—	—	(4,b̸,[illegible]
4	—	—	—	(5,*D*,*L*)	—	—	—	—	—	—
5	—	—	(6,*C*,*L*)	(5,*d*,*L*)	—	—	(5,*C*,*L*)	—	—	—
6	—	(7,*B*,*L*)	(6,*c*,*L*)	—	—	(6,*B*,*L*)	—	—	—	—
7	(8,*A*,*L*)	(7,*b*,*L*)	—	—	(7,*A*,*L*)	—	—	—	—	—
8	(9,*a*,R)	—	—	—	—	—	—	—	(10,$,R)	—
9	—	(9,*b*,R)	(9,*c*,R)	(9,*d*,R)	(9,*A*,R)	(9,*B*,R)	(9,*C*,R)	(4,*D*,*L*)	—	—
10	—	—	—	—	(10,*A*,R)	(10,*B*,R)	(10,*C*,R)	(10,*D*,R)	—	(11,b̸,[illegible]
11	—	**stop and accept**								

For the following we require two facts about Turing machines. The first fact is that for every Turing machine there is an equivalent Turing machine (i.e., one which accepts the same set of words) with exactly one final state k_f. The second one is the fact that it is undecidable whether such a Turing machine when started with a blank tape (i.e., all squares contain b̸; this represents the empty word) will ever enter its final state (thereby accepting the empty word).

Now we can state the main theorem about the decidability of the safety problem for general authorization systems.

Theorem 20. It is undecidable whether a given configuration of a given authorization system is safe for a given generic right.

Proof. As indicated, the idea is to simulate the behavior of an arbitrary Turing machine TM by an authorization system (R, C) in such a way that release of a right corresponds to the Turing machine's entering the final state k_f. Thus let TM $= (K, T, d, k_0, F)$ be a Turing machine with $F = \{k_f\}$. The set R of generic rights will contain a right corresponding to each of the states k in K and to each of the tape symbols t in T; additionally there are the two rights **end** and **own**. In the following we will denote a state or a tape symbol and its corresponding right in R by the same symbol, thus we can write

$$R = K \cup T \cup \{\textbf{end, own}\}$$

Note that K and T do not have any element in common. Before we define the commands c of our authorization system (R, C) we informally describe how the system represents the tape and the moves of TM.

At any given point in time, TM will have scanned some finite initial portion of its tape, say, squares 1 through h. This situation will be represented by a sequence of h subjects $s_1, \ldots, s_h$ such that s_i has the right **own** to the object s_{i+1} for i in $\{1, \ldots, h-1\}$. Thus the **own** relation orders the subjects into a linear list representing the tape of TM (more specifically, the scanned portion of the tape). Subject s_i represents square i, and the content (tape symbol) X of this square i is represented by giving the subject s_i the generic right X to itself. The fact that k is the current state and that the tape head is scanning square j is represented by giving subject s_j (which corresponds to square j) the generic right k to itself. Recall that $K \cap T = \emptyset$; thus we know which right corresponds to a state and which right corresponds to a tape symbol. The generic right **end** marks the last subject s_h. More specifically, we record that s_h is the last subject (the subject representing the square with the highest number which has been scanned so far by TM) by giving s_h the generic right **end** to itself. This also indicates that subject s_{h+1} which s_h is to own eventually (if TM gets that far) has not yet been created. Note that there will always be only one subject with a right k (in K) as TM will be only in one state at any given time.

We now come to the moves of TM and their simulation in (R, C). They are represented by commands of our authorization system. First let us consider what to do if

$$d(k, X) = (p, Y, \text{LEFT})$$

for k and p states, and X and Y tape symbols. Informally there should be two subjects, say, s and s', representing two consecutive tape squares. TM is in state k scanning the square represented by s' which contains the tape symbol X. What is to be achieved by the command is the following: We must take away right k from s', we must change right X into right Y, and we must give subject s the right p. This is exactly what the command c_{kX} does:

command $c_{kX}(s, s')$
  if own in (s, s') **and** $\{k, X\}$ in (s', s')
  then
    delete $\{k, X\}$ from (s', s')
    enter p into (s, s)
    enter Y into (s', s')
  end

If the head is to move to the right, i.e.,

$$d(k, X) = (p, Y, \text{RIGHT})$$

we must distinguish two cases. In the first case the square to the right of the currently scanned one has already been scanned previously; in the second case the tape head moves to a new square which has not been visited before by TM. The first case is completely symmetric to the above:

```
command c_kX(s, s')
    if own in (s, s') and {k, X} in (s, s)
    then
        delete {k, X} from (s, s)
        enter p into (s', s')
        enter Y into (s, s)
    end
```

Note that TM is deterministic; therefore $d(k, X)$ has only one result. Consequently we can use the same name c_{kX} for the command regardless of whether the head moves to the right or to the left.

In the second case we first have to create a new subject as we extend the portion of the tape scanned so far. Note that this situation is detected by the right **end** in the subject; also note that there is precisely one such subject:

```
command e_kX(s, s')
    if {end, k, X} in (s, s)
    then
        delete {end, k, X} from (s, s)
        create subject s'
        enter {b̸, p, end} into (s', s')
        enter Y into (s, s)
        enter own into (s, s')
    end
```

Note that in this case we need a new name (e_{kX}) for the command as the move function d in TM does not distinguish the two situations. Note, however, that c_{kX} and e_{kX} are mutually exclusive.

It follows that in each configuration of the authorization system there is at most one command applicable. Thus (R, C) simulates exactly

TM using the representation as given. If TM enters the state k_f (its only final state) then the authorization system can release the corresponding generic right k_f; otherwise it is safe for the generic right k_f. Since it is undecidable whether TM will ever enter this state, it is also undecidable whether the authorization system is safe for the generic right k_f. □

We conclude this section by discussing some restrictions on these authorization systems. An authorization system is called mono-operational if each command contains only one single primitive operation; it is called monoconditional if there is only one condition after the **if** of the command. For these restricted systems the conclusion of Theorem 20 does not necessarily remain valid anymore. This is illustrated by the following:

Theorem 21. There is an algorithm which decides whether or not for a given mono-operational authorization system a given initial configuration is safe for a given generic right.

However, this assumption is quite restrictive; the model is not very realistic any longer. A similar situation exists if we consider authorization systems without create operations; in this case the safety problem is decidable but we have the following:

Theorem 22. The safety problem for authorization systems without create operations is complete in polynomial space.

Again we know that we can answer the safety problem for a somewhat unrealistic class of systems but even if one was prepared to accept this restriction (which makes the system rather static as there will always be just the objects one starts out with), the theorem tells us that we probably do not want to compute a solution as problems which are complete in polynomial space are considered very hard (it is commonly conjectured that such decision problems require at least exponential time).

While from a practical point of view the last theorem is not very appealing it is interesting to note that it can be extended quite substantially. Instead of imposing the rather drastic assumption that there be no create commands one might consider the question of what happens ie there is only a finite number of subjects; note that this is a very reasonablassumption for any practical authorization system. The result is sumf marized in the following:

Theorem 23. The safety problem for arbitrary authorization systems with a finite number of subjects is decidable.

Let us move from systems which do not create (or only restrictedly) to systems which do not destroy, more specifically systems which do not use the primitive operations

destroy subject s
destroy object o
delete r from (s, o)

Such systems are called monotonic as one can only create but not destroy. It turns out that this restriction does not make much difference; the safety problem remains undecidable.

Theorem 24. It is undecidable whether a given configuration of a given monotonic authorization system is safe for a given generic right.

Thus the ability to remove objects or rights is not important for the safety problem; however, this result is not counterintuitive. The result can be strengthened as stated in the following:

Corollary 6. The safety problem for monotonic authorization systems is undecidable even when each command contains at most two conditions.

However, for monoconditional monotonic systems the situation changes:

Theorem 25. The safety problem for monotonic monoconditional authorization systems is decidable.

This concludes our presentation of the Harrison–Ruzzo–Ullman results and their variants. The kind of authorization systems considered, while certainly general enough to model most currently existing authorization systems, serves primarily to delimit what can be done. As we pointed out in the beginning, one should carefully contemplate their meaning; in particular it would be erroneous to conclude that it is impossible to decide in a particular case whether a certain configuration is safe for a specific right. Furthermore, the statements of Theorems 21, 25, and especially 23 are of a very practical importance as we point out in the following:

Remark. All these algorithms are not useful if a casual user wishes to decide whether or not to give some other casual user rights to a certain file. However, for an important database system which is to be made accessible to a substantial number of people and which contains files with very sensitive information, it is indeed conceivable that a certain overall policy is to be established using one of these "impractical" methods. Note that this is to be done only once in order to guarantee the soundness of this policy. The important point we wish to make here is that no matter how complicated the method, it is to be executed just once for the life time of the database system. Thus in terms of the overall execution time of all the requests to be posed to the system this amount of work, great as it may well be, is only a constant.

Bibliographic Note

The model of an authorization system is taken from Harrison *et al.*[39] (there called a protection system) as are the notions of releasing a right (there called leaking) and of safeness with respect to a certain right. Theorems 20, 21, and 22 are from the same paper. Theorem 23 is from Lipton and Snyder,[61] where a surprising connection between vector addition systems and authorization systems is established. Finally, Theorem 24 and Corollary 6 as well as Theorem 25 are taken from Harrison and Ruzzo.[38] All the proofs not given here may be found in the referenced papers.

Exercises

1. Consider the example on p. 117. In the given configuration (i.e., with $P[\text{COB2}, \text{OS}] = \{e\}$) is there a way whereby COB1 can get COB2 deleted? What if $P[\text{COB1}, \text{COB2}] = \{w\}$?

2. Consider the following example:

$$0 = \{1, 2, 3, 4\}, \ S = 0$$

	1	2	3	4
1	*druw*	*r*	*u*	*w*
2	*w*	*druw*	*r*	*d*
3	*d*	*u*	*druw*	*w*
4	*d*	*u*	*r*	*druw*

Show that in this configuration, 1 can cause 4 to be deleted. Hint—Use the power of update, e.g., if 4 can update 2 then 4 can cause 2 to delete 1 (as 2 has this right).

3. Consider the following Turing machine $M = (K, T, d, k_0, F)$, where $K = \{k_0, k_1, k_2, k_3, k_4\}$, $T = \{x, y, \flat\}$, $F = \{k_4\}$, and the next-move function is given by the table below:

	x	y	$\flat$
k_0	(k_1, y, R)	(k_4, x, R)	(k_3, x, R)
k_1	(k_4, y, L)	(k_3, x, R)	(k_2, y, R)
k_2	(k_0, x, L)	(k_1, y, L)	$(k_2, \flat, L)$
k_3	(k_0, y, L)	(k_0, x, R)	$(k_0, \flat, L)$
k_4	(k_1, x, R)	(k_2, y, R)	$(k_4, \flat, L)$

Determine whether the following words are accepted: x, xy, xx, xxx.

4. Construct an authorization system (R, C) which simulates the Turing machine M given in Exercise 3.

5. In the authorization system (R, C) of Exercise 4, can the right k_4 ever be released? Justify your answer!

6. Give an algorithm to test whether a command c releases a generic right r from a configuration $Q = (S, O, P)$. The parameters for this algorithm should be only (R, C), c, r, and Q. What is the time complexity of this algorithm?

7. Give a procedure which will halt if for a given authorization system (R, C) and a given right r in R, a given configuration Q is unsafe for r. The parameters for this procedure are (R, C), r, and Q.

8. Prove that this procedure can be made into an algorithm (i.e., always halts) if the number of objects is bounded by a fixed constant. This implies that the safety problem is decidable if there is only a finite number of objects. Note, however, that this is a weaker statement than Theorem 23!

3.3. Authorization Systems with Tractable Safety Problem

The last section was devoted to a very abstract theory of authorization systems; we showed that considerably general types of such systems may not have a decidable safety problem or else the algorithms for solving this problem are so time consuming that they can be applied on a limited basis only. In this section we move on to the other extreme; we are interested in authorization systems which have at least a polynomial, preferably a linear time algorithm to answer the safety problem. A large number of such systems have been studied in the last few years; most of them are some variant of an authorization system known by the name "take–grant." This method is of interest as it is the first (and virtually only) system suggested by, and used in, the real world which has been successfully studied. Following Budd and Lipton we will choose a substantially different starting point for our investigation, namely, grammatical authorization systems. These systems have polynomial-time algorithms for the safety problem. An important subclass of them, the so-called regular authorization systems, have linear-time algorithms. Despite the different approach it will turn out that one of these regular grammatical systems is the take–grant system.

3.3.1. Grammatical Authorization Systems

The basis of this section will be the following model of an authorization system. There are n objects in the system, $x_1, \ldots, x_n$, and each object x_i is of type t_i. We will assume that there are finitely many different types t_i; the set of all these types is denoted by T. In the previous section we dealt only with two types, subjects and data objects. Between any two objects x_i and x_j there may be an arbitrary (finite) number, zero or more, of rights. Again we will allow only a finite number of different rights and denote their set by R. For reasons which will become clear soon we will further assume that both T and R are alphabets, i.e., the types t_i and the rights r_i are symbols; a sequence of these symbols can then be considered a word over the appropriate alphabet. (Note that this is no restriction on the generality of the model.) In particular, the empty word (representing a sequence of no rights or types) is denoted by ϵ.

In the last section we represented the status of an authorization system (configuration) by an access matrix. Equivalently we may choose to represent it by a directed graph $G = (V, E)$, where each vertex v in V is some object (together with its type), and whenever there is a right r between objects x_i and x_j there will be a directed edge $(x_i, x_j; r)$ with label r in G. As before, we need rules to transform one graph (configuration) into another. These rules capture the dynamic nature of our systems. Recall that in the last chapter we used commands for this. Their function will be taken over by transition rules in the present setting. Suppose we are in a certain initial configuration and assume that by applying a finite number of these transition rules we can connect vertex x to a vertex y by an edge with label r then we will say that in this initial configuration,

$$x \text{ can-}r\ y$$

Now we can state the definition of the authorization systems we consider in this section. An authorization system is called grammatical if for each right r in R there is a grammar $G_r = (N_r, T_r, P_r, S_r)$ such that given any two vertices (objects) x and y, x and y are connected by a path whose label is in $L(G_r)$, the language generated by grammar G_r, iff x can-r y.

The advantage of having such a grammatical authorization system is that under certain assumptions about the grammars G_r one can obtain very efficient algorithms for solving the safety problem. In fact, in many cases it will turn out that we can use these grammatical authorization systems to get a complete description of the changes produced in the system by giving object x_i the right r to object x_j. This will be referred to as the extended safety problem.

As the notion of a grammatical authorization system is not very intuitive let us first show its feasibility by using it to work with general edge moving systems, another form of authorization systems. A general edge moving system is an authorization system represented by a graph $G = (V, E)$ where the transition or rewriting rules are of the following form. If v in V is of type t_i, for $i = 1, 2, 3$, and $(v_1, v_2; r_2)$, $(v_2, v_3; r_3)$ are in E for rights r_2, r_3 in R, then applying the transition rule yields a graph $G' = (V, E')$, where $E' = E \cup \{(v_1, v_3; r_1)\}$ for some right r_1 in R. Graphically this may be represented by

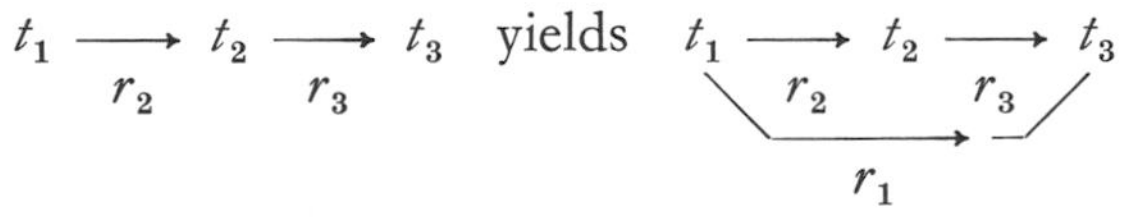

$$t_1 \xrightarrow[r_2]{} t_2 \xrightarrow[r_3]{} t_3 \quad \text{yields} \quad t_1 \xrightarrow[r_2]{} t_2 \xrightarrow[r_3]{} t_3, \quad t_1 \xrightarrow[r_1]{} t_3$$

In particular, an edge moving system never deletes objects. However,

as far as the safety problem is concerned this restriction is not very severe. (Cf. Theorem 24. The ability to destroy is of relevance only because it allows to *hide* the particular way in which a right had been acquired; however, it does not *per se* enable a user to acquire a right!)

Given a general edge moving system we can construct a grammar $G = (\text{VAR}, \text{TER}, \text{PRO}, \text{STA})$ as follows. Whenever there can be an edge with label r (in R) between two vertices (objects) of types t and t' (t, t' in T), there must be a variable (t, r, t') in VAR: thus we will assume without loss of generality that

$$\text{VAR} = T \times R \times T$$

For any variable $A = (t, r, t')$ we create a terminal $a = [t, r, t']$, a in TER. The starting symbol STA will be unspecified; in fact we will consider a class of grammars which differ only in the starting symbol which can range over all variables in VAR. Finally we define the productions as follows: Whenever we have a rewriting rule as above we add the production

$$A \to BC$$

to PRO, where $A = (t_1, r_1, t_3)$, $B = (t_1, r_2, t_2)$, and $C = (t_2, r_3, t_3)$. This production captures precisely the syntactic meaning of the rewriting rule; the three nonterminals encode the rights as well as the types of the objects connected by these rights. Finally we add to the set PRO of productions of the grammar G all productions of the form

$$A \to a$$

where $A = (t, r, t')$ is a variable and $a = [t, r, t']$ is its corresponding terminal. Since R and T are finite so are VAR and TER; PRO is also finite since all productions are of the form $A \to BC$ or of the form $A \to a$ for A, B, C in VAR and a in TER. Such a grammar is called a Chomsky normal form grammar (see, for instance, Hopcroft and Ullman[42]); in particular, these grammars are all context free.

Let us now define the context-free language $L(A)$ to be the language generated by the grammar $G = (\text{VAR}, \text{TER}, \text{PRO}, A)$ with A in VAR. Then we have the following:

Lemma 6. Let two objects v and w of types s and t, respectively, be given. Object v can-r object w iff there exists a path between v and w with a label in the language $L(A)$ for $A = (s, r, t)$.

Proof. If v and w are connected by a path with some label in $L(A)$, then the derivation of this word in the grammar G starting in A gives us a precise description of the sequence of rewriting rules we must apply in order to join v and w by an edge with label r; whenever we use the production $B \rightarrow CD$ in the derivation of the word we apply the corresponding rewriting rule in the graph. Consequently, v can-r w. The converse follows by induction on the number of applications of the transition rules which eventually show that v can-r w. For the basis we let this number be zero, i.e., already in the initial configuration we have v can-r w; thus we trivially obtain the result, as $L((s, r, t))$ clearly contains the word $[s, r, t]$. Therefore let us assume that in the initial configuration v did not have right r to w, and furthermore we assume that n applications of the transition rules were necessary for v to obtain right r to w. Therefore there must exist a vertex u and edges $(v, u; r_2)$, $(u, w; r_3)$ in E_{n-1} and a transition rule which gives v the right r to w in this situation. Here E_{n-1} is the set of edges in the graph after the first $n - 1$ applications. Clearly, to obtain the edge $(v, u; r_2)$ takes less than n applications of the transition rules, similarly for $(u, w; r_3)$. Consequently, by the induction hypothesis, there must be a path between v and u with label z_1 and a path between u and w with label z_2 where z_1 is a word in $L((v, r_2, u))$ and z_2 is a word in $L((u, r_3, w))$. But with the rewriting rule used in the last step the following production in G is associated:

$$(v, r, w) \rightarrow (v, r_2, u)(u, r_3, w)$$

This shows that v and w are connected by a path with a label in $L((v, r, w))$, namely, $z_1 z_2$. □

In view of the results in the last chapter, particularly Theorem 23, the reader should note that no assumption is made about the number of objects involved; only the number of types and rights is fixed. The following result then justifies the use of grammatical authorization systems. Recall that we assume n objects where n is arbitrary.

Theorem 26. The extended safety problem can be answered for a general edge moving system in time $O(n^{2.81})$.

Proof. Assume that the authorization system is represented in an (n, n) matrix M with the entry in row i and column j being the set of rights (in R) which object x_i has to object x_j. Since for technical reasons we need an upper triangular matrix we create for each right r a new right r', the "inverse" right to r, by

$$x \text{ has right } r \text{ to } y \text{ iff } y \text{ has right } r' \text{ to } x$$

Clearly now we can modify our matrix in such a way that all the entries below the main diagonal are empty sets. We contend that this reduces the extended safety problem to the determination of the transitive closure of such a matrix. The closure is with respect to a "matrix multiplication" defined as follows: Addition of scalars in standard matrix multiplication becomes set union, and scalar multiplication becomes the operation of substitution reduction ($BC = A$ iff $A \rightarrow BC$ in PRO). This transitive closure can be found in $O(n^{2.81})$ operations (see Valiant[(95)]). Now, in order to actually solve the extended safety problem we simply determine this closure twice, once without and once with the additional edge, and compare the two results. □

Example. Consider an authorization system with just three rights: read, indirect, and request. Read should be clear; indirect allows to pass on the read right, namely, if x has the indirect right to y and y has read right to z, then x has read right to z; request says that x obtains indirect rights to z if x has request rights to y and y has indirect rights to z. Finally, if x has read rights to y and y has request rights to z then x has request rights to z. The interpretation of this last rule is as follows. Since x can read y and y can request from z it is possible that using x's read right, x can also request from z (see Exercise 2, p. 125). If we write A for read, B for indirect, and C for request, then we have the following grammar corresponding to these rewriting rules:

$$G = (\{A, B, C\}, \{\text{re}, \text{in}, \text{rq}\}, \text{PRO}, \text{STA})$$

with PRO given by

$$A \rightarrow BA \mid \text{re}$$
$$B \rightarrow CB \mid \text{in}$$
$$C \rightarrow AC \mid \text{rq}$$

By Theorem 26 we have a method for solving the safety question. It can be shown that the language generated by this grammar (for any starting symbol, A, B, or C) is not regular. (This follows, for instance, from the fact that

$$(\text{re in rq})^i \ (\text{re rq in})^j \ \text{re}$$

is in $L(A)$ iff $i > j$.)

3.3.2. Regular Authorization Systems

Arbitrary edge moving authorization systems correspond to grammatical authorization systems where the grammar is context free in general. In the last section we saw an example where the language was not regular. This suggests to consider the subclass of regular authorization systems; these are systems where the languages generated by the associated grammars are all regular.

The practicality of this notion will emerge when we show that a very widely studied model and some of its modifications are in fact regular authorization systems. Even more important is the fact that for any regular system there is an algorithm to solve the safety problem which is linear in the size of the graph representing the system, the so-called authorization graph.

Theorem 27. For regular authorization systems the safety problem can be solved in time linear in the size of the authorization graph.

Proof. The central observation is the fact that for a regular grammar there is a finite automaton recognizing the language generated by the grammar. Note that the regular grammar and therefore the corresponding finite automaton do not depend on the size of the graph; consequently, in terms of n, the number of objects (= the number of vertices in the graph), the automaton has a constant number of states, say, c. We now construct a new graph G' with cn vertices by putting an edge from (x_i, q) to (x_j, q') iff there is an edge between the objects x_i and x_j in the authorization graph with label r and the result of the state q of the automaton under r is the state q'. Let x now be the object for which we want to solve the safety problem, namely, we want to determine what other objects can obtain the right r to y if we give x the right r to y. Starting from the vertex x in the original authorization graph G we can construct the new graph G' by depth-first search in $O(e)$ operations where e is the number of edges in G. Furthermore, in the process of constructing G' we mark those vertices which we encounter while the finite automaton is in a final state. It should be clear that the marked vertices are precisely those vertices in G' (objects of the system) which can obtain rights to x (by definition of a grammatical authorization system). □

Clearly, in view of this result regular authorization systems are very desirable. However, it is well known (see, for instance, Hopcroft and Ullman[42]) that given an arbitrary context-free grammar it is undecidable whether it generates a regular language; in fact, this also holds for grammars in Chomsky normal form. Thus instead of vainly searching for a characterization of regular authorization systems which cannot exist (at least not an effective one), let us consider a large class of regular authorization systems. This leads to the notion of a nondiscriminating grammar.

Informally an authorization system is nondiscriminating if all the transition rules are of the following form:

If object x has a certain right r to object y, and y has *any* right r' to object z, then x can obtain this right r' to z.

Thus the rules do not discriminate between different rights on the second edge. Formally, an authorization system is nondiscriminating if it has a nondiscriminating grammar. A nondiscriminating grammar $G =$ (VAR, TER, PRO, STA) is defined below; note that this *grammar* is not regular. The set VAR of variables consists of five different types of variables, namely,

A_h for h in $\{1, \ldots, H\}$,
B_i for i in $\{1, \ldots, I\}$,
C_j for j in $\{1, \ldots, J\}$,
D_k for k in $\{1, \ldots, K\}$, and
the single variable Z.

TER is an arbitrary alphabet of terminals, and STA, the starting state, is unspecified for the time being. Finally we must define the set PRO of productions:

(1) For every variable X, X being one of the A_h, B_i, C_j, or D_k, there is a production of the form $X \rightarrow w$ where w depends on X and is a single word over the alphabet TER of terminals.

(2) In addition to the above, some subset of the total set of productions, listed below, is in PRO (but not necessarily all of them!):

(a) $A_h \rightarrow w_{h,h'} A_{h'}$ for all h, h' and all $w_{h,h'}$ words over TER.

(b) $B_i \rightarrow B_{i'} v_{i,i'}$ for all i, i' and all $v_{i,i'}$ words over TER.

(c) $C_j \rightarrow ZA_h$, $C_j \rightarrow ZB_i$,
$C_j \rightarrow A_h$, $C_j \rightarrow B_i$,
for all h, i, and j.

(d) $D_k \to A_hZ$, $D_k \to B_iZ$,
$D_k \to A_h$, $D_k \to B_i$,
for all h, i, and k.

(e) $Z \to A_h$, $Z \to B_i$, $Z \to C_j$, $Z \to D_k$,
$Z \to C_jD_k$, $Z \to ZZ$, $Z \to \cent$
for all h, j, and k.
($\cent$ denotes the empty word.)

While these nondiscriminating grammars are not regular it turns out that they generate regular languages as stated in the following:

Theorem 28. The language generated by an arbitrary nondiscriminating grammar G, with STA being some variable in VAR, is regular.

Proof. First it is clear that any A_h (any B_i) produces some regular language, say, $L_{A_h}(L_{B_i})$. Consequently, if we can show that Z produces a regular language the claim follows. In the following we will construct a regular expression for the language generated by (VAR, TER, PRO, Z).

(a) First we assume there are none of the productions $Z \to C_jD_k$ nor the production $Z \to ZZ$ in PRO. Let us define a regular language of terminal words L (for left) as follows:

- If there are productions $Z \to D_k$ and $D_k \to A_hZ$ in PRO, then L_{A_h} is contained in L.
- If there are productions $Z \to D_k$ and $D_k \to B_iZ$ in PRO, then L_{B_i} is contained in L.
- No other words are in L.

Similarly we define a regular language R (for right) as follows:

- If there are productions $Z \to C_j$ and $C_j \to ZA_h$ in PRO, then L_{A_h} is contained in R.
- If there are productions $Z \to C_j$ and $C_j \to ZB_i$ in PRO, then L_{B_i} is contained in R.
- No other words are in R.

Finally the regular language F is defined as follows:

- For all X, X being one of the A_h, B_i, C_j, or D_k, if there is a production $Z \to X$ in PRO then L_X is contained in F.
- No other words are in F.

Thus the set of words over TER $\cup \{Z\}$ we can generate in this case from Z is

$$L^*(Z \cup F)R^*$$

(b) Now assume we add the production $Z \to ZZ$. It is trivially verified that in this case we can generate exactly MM^* if we can generate M without this production. Consequently if we add the production $Z \to ¢$ we get

$$(L^*(Z \cup F)R^*)^*$$

Let us denote this expression in the following by $[LR]$.

(c) Finally we have to add the productions of the form

$$Z \to C_j D_k$$

Define W to be the union of all languages

$$L_X \cdot L_Y$$

where $C_j D_k$ derives $ZXYZ$ using productions $C_j \to ZX$ and $D_k \to YZ$, for X some L_{A_h}, Y some L_{B_i}. Clearly W is again regular. It is not difficult to see that with one application of a production of the form $Z \to C_j D_k$ one can produce precisely

$$[LR]W[LR]$$

A tedious but straightforward argument shows then by induction on the number of applications of these productions that Z can produce

$$([LR]W)^*[LR]$$

using arbitrarily many applications of these productions. Now we finally substitute $¢$ for all Z in $[LR]$ as everything else has already been taken care of; this yields

$$((L^*(¢ \cup F)R^*)^*W)^*(L^*(¢ \cup F)R^*)^*$$

It can be seen that this is equivalent to the regular expression

$$(L \cup F \cup R \cup W)^*$$

Note that in a sense this is a "maximal" expression, as some of the productions used may not have been present to start with. In such a case, however, it should be clear from this general derivation how to obtain the corresponding regular expression. □

It follows that any such system allows to solve the safety problem in linear time.

Corollary 7. Any nondiscriminating authorization system has an algorithm for answering the safety problem in time linear in the size of the authorization graph.

The proof is an immediate consequence of Theorems 27 and 28.

Exercises

1. Consider the following rules in an edge moving system:

   (a) If x has right r to y, and y has right r to z, then x can acquire right r to z, or shorter,

   $$x \text{ can-}r\ y \text{ and } y \text{ can-}r\ z \text{ imply } x \text{ can-}r\ z.$$

   (b) x can-w y and y can-w z imply x can-w z.

   (c) x can-r y and z can-w y imply x can-r z.

   Determine the grammars corresponding to these rules, assuming that all objects are of the same type. Are the languages L(STA) regular for STA ranging over all variables?

2. Consider the same edge moving system as in Exercise 1; however, now assume that there are two types t_1 and t_2 and the first two rules hold only if x, y, and z have the same type, while the third rule holds only if x and z have the same type, which in turn must be different from the type of y. Determine the grammars corresponding to these rules. Are the languages L(STA) regular for STA over all variables?

3. Consider the edge moving system discussed in Exercise 2 but *add* the following rule:

   (d) If x and y are of type t_1 and z is of type t_2, then x can-r y and y can-r z implies z can-r z $(t_1 \neq t_2)$.

   Determine the grammars corresponding to these rules. Are any of the languages L(STA) not regular?

3.3.3. An Application: The Take–Grant System

As an application of grammatical, and in particular regular, authorization systems we will consider the take–grant system in its more realistic version which distinguishes between subjects and objects. This system and several variants are studied extensively in the literature. Moreover, they are of practical interest; indeed the system was first proposed by practitioners and not by theoreticians.

We will first briefly describe this system in terms of edge moving systems and then give a nondiscriminating grammar for it thereby proving immediately (by Corollary 7) that the subject/object take–grant system has a safety problem solvable in linear time.

In this model each vertex is of one of two types, subject (S) or object (O). There are precisely two rights, read (r) and write (w); thus an edge may have one of the three labels r, w, or rw. There are three transition rules: take, grant, and create. If we do not say anything about the type of vertex it may be S or O. The right s can be r, w, or rw.

(a) Take: Let x be a subject having right r to y, and let y have right s to z. By the take rule, x acquires right s to z.

(b) Grant: Let x be a subject having right w to y and right s to z. By the grant rule, y acquires right s to z.

(c) Create: Let x be a subject. By the create rule x may create a new subject to which x then has the rights rw.

This system can be shown to have a nondiscriminating grammar $G = (\text{VAR}, \text{TER}, \text{PRO}, \text{STA})$ defined as follows:

$$\text{VAR} = \{S, O\} \times \{r, w, r', w'\} \times \{S, O\} \cup Z$$

where S and O are the two types, r and w are the standard rights, and r' and w' are the inverse rights given by

$$x \text{ has right } r\ (w) \text{ to } y \text{ iff } y \text{ has right } r'\ (w') \text{ to } x.$$

For brevity we use the following abbreviations:

$A = (S, r, S)$ $\quad B = (S, r, O)$ $\quad C = (O, r, S)$ $\quad D = (O, r, O)$

$E = (S, r', S)$ $\quad F = (S, r', O)$ $\quad G = (O, r', S)$ $\quad H = (O, r', O)$

$I = (S, w, S)$ $\quad J = (S, w, O)$ $\quad K = (O, w, S)$ $\quad L = (O, w, O)$

$M = (S, w', S)$ $\quad N = (S, w', O)$ $\quad O = (O, w', S)$ $\quad P = (O, w', O)$

For all these variables (A through P) there is a corresponding terminal in TER:

$$\mathrm{TER} = \{a, b, \ldots, p\}$$

The variable A will be the starting symbol STA. Finally, the set PRO of productions. For brevity we will use a self-explanatory shorthand. For each of the variables A through P there is one of these "productions," and for Z there are eight:

$$A \to Z(a \cup bd^*c)$$
$$B \to Zbd^*$$
$$C \to OA$$
$$D \to OB$$
$$E \to (e \cup fh^*g)Z$$
$$F \to EJ$$
$$G \to h^*gZ$$
$$H \to GJ$$
$$I \to Z(i \cup bd^*k)$$
$$J \to Z(j \cup bd^*l)$$
$$K \to OI$$
$$L \to OJ$$
$$M \to (m \cup nh^*g)Z$$
$$N \to MJ$$
$$O \to (o \cup ph^*g)Z$$
$$P \to OJ$$

$$Z \to ZZ \quad Z \to A \quad Z \to E \quad Z \to I \quad Z \to M$$
$$Z \to JG \quad Z \to BO \quad Z \to ¢$$

With A being the starting symbol we can remove the indented productions for C, D, F, H, K, L, N, and P; this leaves us with a nondiscriminating grammar. It should be clear how these abbreviations are to be handled, e.g., instead of writing $A \to Z(a \cup bd^*c)$ one would write $A \to ZY$, $Y \to a$, $Y \to Y''c$, $Y'' \to Y''d$, $Y'' \to b$. Following the proof of Theorem 28 one then derives a regular expression associated with A:

$$(a \cup bd^*c \cup e \cup fh^*g \cup i \cup bd^*k \cup m \cup nh^*g$$
$$\cup\, bd^*o \cup bd^*ph^*g \cup jh^*g \cup bd^*lh^*g)^*$$
$$\cdot\, (a \cup bd^*c)$$

Now Theorem 27 gives us a convenient method to solve the safety problem for the subject/object take–grant system in time linear in the number of subjects and objects.

Bibliographic Note

The take–grant system originated in practical systems rather than in theoretical studies; it was proposed in Graham and Denning[34] and in Jones.[43] It subsequently was analyzed in several papers, notably in Lipton and Snyder[60] for subjects only and in Jones *et al.*[46] for subjects and objects. A large number of related papers was spawned by these publications dealing with extensions and modifications of the take–grant system as well as its critique from a practical point of view. Here we mention Snyder,[91,92] and Bishop and Snyder,[5] and for a critique, Jones.[45] An attempt at defining useful classes of authorization systems and thereby unifying the theory of such systems with tractable safety problem are grammatical authorization systems which were introduced in Lipton and Budd;[59] most of the material in this section is from this paper. Furthermore, it presents an extension, namely, "near-grammatical systems" which also have a linear-time solvable safety problem.

3.4. Overview of a Practical Implementation

In the last two sections we were primarily concerned with theoretical aspects of authorization systems. In this section we will give a brief overview of an authorization system which is actually implemented in a real-world environment. In the introduction to this topic we will also touch upon the motivation for decentralized data management and the case against monolithic databases with a central administrator, as this subject is intimately related to data security. This is an area which is in transition due to the relatively recent introduction and acceptance of distributed data management and distributed processing in the commercial field. Then we describe the authorization system which is implemented in IBM's System R, a project of a commercially useful relational database system.

3.4.1. Motivation

In this section we will discuss the need for an authorization system from a very practical point of view. The central buzz-word is "sharing".

All the problems related to authorization would be entirely irrelevant if nothing was to be shared. Clearly with very few exceptions (home computer, for example) sharing is mandatory. Almost all computer installations have more than one user at the same time, thus the (physical) computer is shared. Some of these users have more "authority" than others (nobody is very happy when a novice programmer accidentally modifies the operating system). Many databases contain information which is of (rightful) interest to several users; they usually can be shared, too. Certain programs must in fact be shared in order to avoid completely unfeasible duplications (for instance, compilers or subroutine libraries).

Most of these ideas are widely accepted; their implementation, however, is almost exclusively based on a central processor having final authority over any action performed in the system. The universal wisdom of this design decision is becoming more and more questionable. The main reason for this we believe to be the following.

The literature is replete with stories about supposedly secure operating systems which were subverted with little effort. It is also generally accepted that the larger and more complex a system, the more things will go wrong. Given these two premises it does not make for very reasonable policy to entrust the security of every piece of information and every program to one huge piece of software, as a subversion of this program guarantees almost unrestricted and often even undetectable (traceless) access to all resources in the system.

Other points in our case against centralized data management are more related to the psychology of the users than to technical questions. Most users who write programs and store them in files would consider themselves owners of their files. While this is natural and desirable, it is nevertheless not true; in virtually all practical systems the operating system is the owner. It merely grants the users permission to access "their" files, more or less gracefully. Unfortunately, many people are brought up to believe otherwise. Proof of this can be gleaned from the fact that the phrase "The system just wiped out my file" is considered a colloquialism; in reality it aptly reflects who is in charge. In other words, if the system had not had access authorization it could not have erased the file. Another point we wish to make is the fact that despite interaction and sharing between users many times the responsibility for the shared object rests with one user only.

All these arguments against centralized data management drastically increase their importance when we move to a database–data communications network. This is an environment where not only access to objects is shared but where different parts of one object may be under the supervision (authority) of different users. A typical example is a large financial institution with many branches which has a database of all the customers and the information pertaining to them. Two approaches are obvious: to have the database stored in one location with one database manager, or to have each branch throughout the country administer its own clients and make these files accessible to the other branches if the need arises. The advantages of the second approach, namely, locally administered databases, should be clear. It is more likely that a client uses the branch where his/her account is kept rather than a different branch. Thus in the first scenario every time any customer makes a transaction the branch must communicate with the central database, whereas in the second scenario this is necessary only for out-of-town customers. Also note that the "solution" of duplicating would create a multitude of problems related to the maintenance of the database (updating, etc.).

These are some of the reasons why a discussion of authorization is very necessary especially if one is primarily interested in the practical implications. The trend definitely is toward database–data communication networks and distributed processing and away from centralized data management. While it is probably true that some of our objections can be overcome in a centralized system our main contention appears unassailable, namely, that it is not very reasonable from a security point of view to entrust one program with access to all sensitive information in the entire system.

3.4.2. The Model

In the last section we made the case for authorization systems from a practical point of view. Now we turn to a specific implementation. We will first present our model. The reader should keep in mind that we are primarily concerned with the sharing of information. Thus the creation of objects is of lesser concern.

Objects come in three kinds: data objects, communication objects, and transaction objects. Data objects are subdivided into physical files and logical files or views. A physical file may be a program or a list of data elements, in short what is usually understood by the word file. A view or logical file is based upon one or more physical file(s) and defines a certain way of looking at the file(s). For example, a view can be used to let a user see only certain fields but not the complete file;

thus a user of a census file may be able to see the field with the number of members of a given family but not the field containing the combined income of the family. However, views do not only act as windows to a file; they can also compute etc. For instance, another user of the census files may be able to get a view which shows the combined income divided by the number of family members without revealing either of the two fields of the physical file. (Note that this is a dangerous example; if the two users exchange information they can in fact determine the combined income.) An important fact about views is that they are not stored in the database system, only their definition is. The reason for this seemingly inefficient procedure—clearly it must be recomputed every time it is used—is the desire that a view reflect changes in a file instantaneously. Thus it is unavoidable to recompute the view. Views are also distinguished from files in the way they are authorized. A definer of a physical file always has all rights to this file as a matter of policy. A definer of a view does not, as clearly these rights depend on the rights the user has to the underlying file(s). Thus, as views can be applied to different files, the authorization of the same view may change.

Communication objects, the second type of objects involved, are logical ports and message queues. They are not central to our considerations; it suffices to assume that they exist and that they function properly. This is not an unreasonable assumption as they are relatively simple components with clearly defined behavior.

Transaction objects, finally, are parameterized programs which perform operations on data objects as well as on communication objects. For example, if we wish to send a file to another user in a network we will do this using some protocol. This operation can be done by employing such a transaction object. Again, they are of lesser importance to us here.

Usage of these objects is governed by the following rules. Each user can use those and only those objects to which s/he has (proper) authorization (permission). Each time a user defined (creates) a new object, this user implicitly holds all rights to this object. Objects can only be shared if the owner grants permission; thus the model is entirely owner oriented. Presumably such a grant will be issued upon request from some other user; however, the transmission of, and action upon, these requests is not part of the database authorization system proper. While it certainly makes sense to grant certain permissions to an unspecified class of users (e.g., the FORTRAN subroutine library should be accessible to any user with a FORTRAN program), this is presently not explicitly implemented. One way of achieving this is by testing all users against a given definition (characterization) of a class of users and giving them the appropriate authorization. The only additional feature required in this case would be access to the file of all users; clearly such a file will exist at any rate. However, this may not be very efficient if the number of users is large. Other methods are possible and have been implemented. [See, for instance, the Digital Equipment Corporation (DEC) systems.]

We distinguish the following authorizations. For data objects (and these are the objects we are primarily interested in) there are several kinds of authorizations, traditional ones and nontraditional ones. Traditional authorizations are those granting rights to specific files, or specific records, or specific fields. These rights include read, write, insert, delete. All these rights also apply to views. Furthermore one would want to have value dependent authorization, that is, access only to those records whose fields satisfy certain conditions. Another nontraditional authorization is that to define virtual files or views, dynamic windows which are not physically stored but synthesized (at request time) from existing (physical) files.

For transaction objects there is another authorization in addition to all the above, namely, that to run (execute) a transaction object.

All authorizations discussed so far are passive, i.e., user A who holds a right allows user B to exercise this right, e.g., to run or update or read. However, every right can be granted in two versions, with or without grant-option. To grant a right with grant-option means that the user to whom the right was granted now is also free to grant the right to somebody (anybody) else while retaining the right him/herself. To grant a right without grant-option removes this possibility of passing on the right. In practical terms if a right is granted with grant-option it is more transparent that the object is now in the public domain, for even if the right was granted to a user without the grant-option there is nothing which will prevent this user from passing on the information even if passing on the right itself is impossible. Clearly this objection does not apply to all authorizations; for example, the execute privilege is useful only if the user can actually exercise it.

This concludes our list of permissions. However, there is one possibility which has not been mentioned at all, namely, the option of revoking rights previously granted. Being able to revoke rights is of considerable practical interest especially in an owner-oriented system like the present one. It does introduce, however, a number of complications, some of which we will discuss later on.

3.4.3. Implementation Considerations

We will now give an overview of some of the more important issues in the implementation of this model. In all that follows it is implicitly assumed that all rights or privileges originate with the creator (definer) of the data object they refer to. The reader should also keep in mind that the implementation assumes a relational database; thence the use of relations, tables, tuples, rows, columns, etc. For the casual reader it suffices to think of the kind of databases discussed in Sections 2.1.1 or 2.2.2. Row and column have then precisely the intuitive meaning, tuple is a synonym for row, and relation or table is a complete database. For a precise definition we refer to the literature (see, for instance, Date[16] or the papers by Codd[11–15]).

A table will be used to represent the privileges pertaining to files or views. As transaction objects do not differ substantially from files for the purposes of representing privileges we will restrict our attention to the following list of privileges which may be stored in a table:

- READ: Can use this relation in a query, permits reading tuples from the relation and defining views based on the relation.
- INSERT: Can insert new rows (tuples) in the relation.
- DELETE: Can delete rows.
- UPDATE: Can modify data in the table; may be restricted to a subset of the columns of the table.
- DROP: Can delete the whole table from the system.

As an example let us assume that user A is the creator of the relation EMPLOYEE, and let us suppose that the following commands were issued:

```
A: GRANT READ, INSERT ON EMPLOYEE TO B WITH
                      GRANT-OPTION
A: GRANT READ         ON EMPLOYEE TO X WITH
                      GRANT-OPTION
B: GRANT READ, INSERT ON EMPLOYEE TO X
```

Note that all commands are legal, i.e., the issuer has the proper authorization to issue the grants; otherwise the grants would be ignored. After executing these commands user X has the right READ with grant-option (from B). Also, when revoking, it will be necessary to remember that X also has READ privilege without grant-option on EMPLOYEE from B, for it is conceivable that A revokes X's right to read EMPLOYEE without revoking B's right to do so. In this case X will still have the READ right passed on from B (however, without grant-option).

The implementation of the model up to now is as follows. We have two relations for the authorization subsystem, SYSAUTH and SYSCOLAUTH. The relation SYSAUTH consists of

USERID TNAME TYPE UPDATE GRANTOP PRIV

which have the following meanings:

TNAME is the name of the table (relation) to which this tuple in SYSAUTH corresponds.

USERID is the user who is authorized to perform the actions recorded in PRIV on the relation TNAME.

TYPE indicates whether TNAME refers to a file or base relation (TYPE = R) or to a view (TYPE = V).

UPDATE indicates whether the user USERID may perform this action on all columns of the relation (UPDATE = ALL), on some columns, (UPDATE = SOME), or on none (UPDATE = NONE).

GRANTOP indicates whether the privileges are grantable or not.

PRIV is a sequence of columns, one for each privilege, which indicate (with Y or N) whether the corresponding privilege may be exercised or not.

This concludes the definition of SYSAUTH. Note that for each (user, table) pair there are zero, one or two entries in SYSAUTH: zero if there are no privileges, one if the privileges are either with or without grant-option, and two if there are some privileges with and some without grant-option, in which case there is one tuple for all privileges with grant-option and one tuple for all without. The relation SYSCOLAUTH is needed only if UPDATE = SOME, meaning that there are some columns on which the user USERID may exercise the privileges and there are others on which s/he may not. SYSCOLAUTH is used to record precisely on which columns that may be done: For each updatable column, a tuple

(user, table, column, grantor, grantopt)

is in SYSCOLAUTH.

Every time a GRANT command is issued, the authorization module verifies whether the grant is authorized. If it is, either a new tuple is inserted in SYSAUTH or an old one is modified; the action on SYSCOLAUTH is according to the UPDATE field of the corresponding SYSAUTH tuple.

Revocation

Any user who has granted a privilege may subsequently revoke it (possibly partially). Now two tuples in SYSAUTH are no longer sufficient, as we must record who granted what to whom. This requires up to two tuples in SYSAUTH for each triple:

(grantee, table, grantor)

Furthermore if a grantor revokes a privilege from some grantee the system must revoke all authorizations which originated in the (now revoked) right.

Recall our example. Suppose the following command is issued by A:

A: REVOKE READ ON EMPLOYEE FROM B

then not only must B's privilege be revoked but also X's READ privilege on EMPLOYEE. Another example is

```
A:  GRANT READ, INSERT, UPDATE  ON EMPLOYEE
                                TO X
B:  GRANT READ, UPDATE          ON EMPLOYEE
                                TO X
A:  REVOKE INSERT, UPDATE       ON EMPLOYEE
                                FROM X
```

Then after execution of these commands, X will still retain the rights READ and UPDATE to EMPLOYEE (provided B's right to grant to X did not originate in a grant by X).

Formally the semantics of REVOKE are defined as follows. Suppose

$$s = G_1, \ldots, G_{i-1}, G_i, G_{i+1}, \ldots, G_n$$

is a sequence of grants where each G_j is a grant of a single privilege. Furthermore assume they are written in the order in which they were issued i.e.,

$$G_j \text{ occurred later than } G_i \text{ iff } i < j$$

Suppose G_i is revoked by R_i; then the meaning of the sequence of grants s followed by the revocation of G_i,

$$s, R_i$$

is defined to be

$$G_1, \ldots, G_{i-1}, G_{i+1}, \ldots, G_n$$

However, in the second example we made an important assumption, namely, that B's privilege to grant READ and UPDATE had not originated in X's privilege. The following problem arises: We must determine whether the revocation affects all the grants or whether some are not affected and can still be retained. This is illustrated by the following examples:

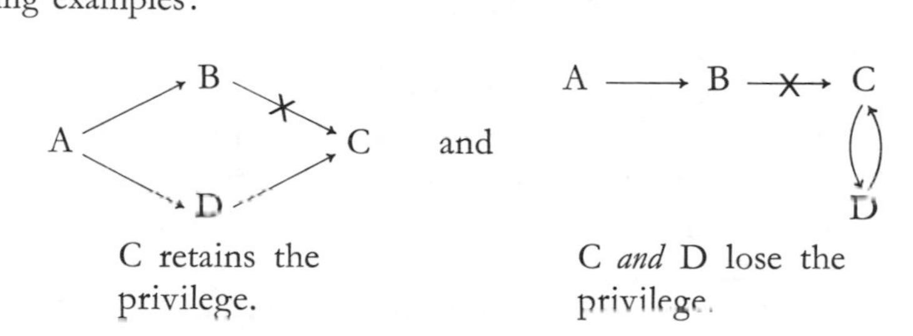

C retains the privilege.

C *and* D lose the privilege.

In order to be able to decide what to do we again modify SYSAUTH by writing in the column for privileges not just Y or N but a time stamp such that the temporal order of the commands is reflected.

Suppose we have in SYSAUTH

USER	TABLE	GRANTOR	READ	INSERT	DELETE
X	EMPLOYEE	A	15	15	0
X	EMPLOYEE	B	20	0	20
X	EMPLOYEE	X	25	25	25
X	EMPLOYEE	C	30	0	30

Suppose at time 35 the command

B: REVOKE ALL RIGHTS ON EMPLOYEE FROM X

is issued. Thus the (X, EMPLOYEE, B) tuple must be deleted from SYSAUTH. Let us determine which of X's grants to EMPLOYEE must be also deleted. We have a list of X's remaining grants and a list of X's grants to others. The grants of DELETE by X at time 25 must be revoked because the earliest remaining DELETE privilege was received at time 30, and thus X used the grant from B at time 20 to grant DELETE to Y. The other two grants, READ and INSERT, remain as they are supported by earlier grants (time 15).

One final modification of this scheme is necessary. Recall that so far granting one privilege twice at different times results in the recording of the earlier grant only. It turns out that there are some situations in which this leads to counterintuitive results. This can be seen as follows. Since one does not record a repetition of an earlier grant, according to the semantics of REVOKE one might remove a grant with a very early time stamp, while in a version where all grants are recorded the later grant would have been removed and the earlier one would have remained untouched.

It is clear that checking of authorization must be done at run time. Also one observes that revocation is fairly expensive in the present scheme. This is essentially a consequence of the first observation.

Authorization on views is not as simple as for relations. The creator of a relation is fully and solely authorized to perform actions on the relation; this is implicit in the definition of a physical relation. The creator of a view, on the other hand, is solely, but not fully authorized because views are defined on top of relations and thus the authorization of a user's superimposed view depends on that user's authorization on the underlying physical relation.

Example. We illustrate this section with an extensive example. For simplicity we number the commands in chronological order and take this numbering as the time at which the commands were issued (time stamp). We will assume that the file F was created beforehand by A; recall that the creator of an object holds all privileges to it. Furthermore we will use the following abbreviations:

READ	R
INSERT	I
DELETE	D
UPDATE	U
DROP	X
WITH GRANT OPTION	WG.

Suppose the following commands are issued:

Time	Grantor	Grants				Grantee	
1	A :	GRANT	RIU	ON F	TO	B	WG
2	B :	GRANT	RI	ON F	TO	C	WG
3	B :	GRANT	RU	ON F	TO	D	WG
4	A :	GRANT	ID	ON F	TO	D	WG
5	B :	GRANT	RU	ON F	TO	Y	WG
6	D :	GRANT	IUD	ON F	TO	C	WG
7	C :	GRANT	RIUD	ON F	TO	Y	WG
8	Y :	GRANT	U	ON F	TO	Z	
9	C :	GRANT	I	ON F	TO	Z	WG
10	D :	GRANT	RIUD	ON F	TO	Z	

It can be verified that all grants are legal, i.e., all grantors have the proper authorization necessary for the commands issued. The relation SYSAUTH at this point (immediately after time 10) is given below; note that SYSCOLAUTH is not required as UPDATE was not restricted to certain columns of F:

USER	TABLE	TYPE	UPDATE	WG	PRIVS	TIME
A	F	R	ALL	Y	RIDUX	0
B	F	R	ALL	Y	RUI	1
C	F	R	NONE	Y	RI	2
D	F	R	ALL	Y	RU	3
D	F	R	NONE	Y	ID	4
Y	F	R	ALL	Y	RU	5
C	F	R	ALL	Y	IUD	6
Y	F	R	ALL	Y	RIUD	7
Z	F	R	ALL	N	U	8
Z	F	R	NONE	Y	I	9
Z	F	R	ALL	N	RIUD	10

Because of the last modification made to conform revocation to our intuitive notion, two entries in SYSAUTH per (user, table)-pair do not suffice any longer as the time of the grant must also be recorded. This is illustrated by the three rows for user Z. Suppose at time 11, A issues the following command:

11 A : REVOKE I ON F FROM B

Clearly this revocation affects some of the other users as well, since some of their privileges originated in A's grant of INSERT to B. After the revocation is performed correctly, SYSAUTH will be as follows:

USER	TABLE	TYPE	UPDATE	WG	PRIVS	TIME
A	F	R	ALL	Y	RIDUX	0
B	F	R	ALL	Y	RU	1
C	F	R	NONE	Y	R	2
D	F	R	ALL	Y	RU	3
D	F	R	NONE	Y	ID	4
Y	F	R	ALL	Y	RU	5
C	F	R	ALL	Y	IUD	6
Y	F	R	ALL	Y	RIUD	7
Z	F	R	ALL	N	U	8
Z	F	R	NONE	Y	I	9
Z	F	R	ALL	N	RIUD	10

In other words, only B actually loses I on F; while the grant of I from B to C (at time 2) is removed, C acquires I from D at time 6 and this grant is not revoked. Consequently C's grants of I to Y and Z remain unaffected. Note, however, that this would not be the case if the grant of I from D to C had occurred *after* C had issued the grants to Y and Z, for instance at time 12. In this case, Y and Z would lose I on F.

Assume the next command issued is

12 B : REVOKE U ON F FROM D

Again this revocation affects several previous grants. The table SYS-AUTH immediately after time 12 is as follows:

USER	TABLE	TYPE	UPDATE	WG	PRIVS	TIME
A	F	R	ALL	Y	RIDUX	0
B	F	R	ALL	Y	RU	1
C	F	R	NONE	Y	R	2
D	F	R	NONE	Y	R	3
D	F	R	NONE	Y	ID	4
Y	F	R	ALL	Y	RU	5
C	F	R	NONE	Y	ID	6
Y	F	R	NONE	Y	RID	7
Z	F	R	ALL	N	U	8
Z	F	R	NONE	Y	I	9
Z	F	R	NONE	N	RID	10

Specifically, the grants of U from B to D (time 3), from D to C (time 6), from C to Y (time 7), and from D to Z (time 10) are removed. However, as Y received U from B at time 5, Y as well as Z (grant of U from Y at time 8) continue to hold the privilege UPDATE on the file F at time 12.

13 D : REVOKE I ON F FROM C

This revocation affects the grants of I from D to C (time 6), from C to Y (time 7), and from C to Z (time 9); however, Z continues holding I on F by virtue of the grant of I from D to Z at time 10. Thus the final SYSAUTH is as follows:

USER	TABLE	TYPE	UPDATE	WG	PRIVS	TIME
A	F	R	ALL	Y	RIDUX	0
B	F	R	ALL	Y	RU	1
C	F	R	NONE	Y	R	2
D	F	R	NONE	Y	R	3
D	F	R	NONE	Y	ID	4
Y	F	R	ALL	Y	RU	5
C	F	R	NONE	Y	D	6
Y	F	R	NONE	Y	RD	7
Z	F	R	ALL	N	U	8
Z	F	R	NONE	N	RID	10

Clearly all revocations can be done recursively. In general it is not necessary to examine all SYSAUTH entries to determine which grants must be removed, provided the appropriate data structures are chosen for the implementation. (This applies to relational databases as well as to conventional databases.)

Exercises

1. Give an outline of the data structures which you would use if you were to implement this authorization mechanism using data structures (conventional databases, i.e., hierarchial or network) instead of relational databases!

2. Describe a recursive revocation algorithm (a) if the implementation is done using relational databases, (b) if the implementation is done using conventional databases. Estimate the time complexity of your algorithms. (It should be clear that using present-day technology relational databases must still be implemented in the conventional way; however, for the user they may be more convenient to use. Thus it is reasonable to write a revocation procedure for the relational implementation provided that this implementation is realized by the system using conventional data structures. The advent of VLSI technology which will allow large-scale parallel processing of information may invalidate this assumption.)

3. Let A be the creator of the file F. Suppose the following sequence of commands has been issued:

1	A :	GRANT RIU	ON F TO	B	WG
2	A :	GRANT RID	ON F TO	C	WG
3	B :	GRANT RI	ON F TO	D	WG
4	B :	GRANT U	ON F TO	C	WG
5	C :	GRANT DI	ON F TO	D	WG
6	C :	GRANT RU	ON F TO	E	WG
7	C :	GRANT R	ON F TO	X	WG
8	D :	GRANT I	ON F TO	X	
9	E :	GRANT U	ON F TO	X	WG
10	X :	GRANT RU	ON F TO	Y	

(a) Describe precisely what privileges X has after time 10. What are Y's privileges at this time?

(b) Suppose at time 11 the following command is issued:

A : REVOKE R, I ON F FROM C

What privileges will have X and Y immediately after time 11?

(c) Suppose at time 12 an additional command is issued:

B : REVOKE U ON F FROM C

What are now the privileges of X and Y?

Bibliographic Note

In Chamberlin *et al.*[9] the motivation for database–data communication systems is discussed at length. Sections 2 and 3 are primarily based on Griffiths and Wade.[35] This paper was amended and extended in Fagin,[30] which contains an example where one intuitively expects the scheme to operate as if all grants were recorded and not just the earliest one. Jones[44] discusses the problem within a much wider context. Stonebraker and Rubinstein[94] describe another implementation of an authorization system. System R is the relational database implementation which also accommodates the authorization system described in this chapter. More information about System R can be found in Astrahan *et al.*[2] For more information about relational databases the reader is referred to any textbook on databases (e.g., Date[16]) or the papers by Codd.[11–15]

3.4.4 Bounded Propagation of Privileges

In this section we will describe a useful extension of the model outlined above. The motivation for this extension is given below.

It is not difficult to see that the model as given in the previous sections does not have a decidable safety problem, i.e., there is no generally applicable algorithm to answer the question whether a given subject can acquire a certain privilege to some object. While this is certainly not an advantage, most users are probably not too concerned about it. They may, however, be more interested in the following question: How far can privileges propagate? More specifically, if A is the creator of F, A will have a legimate interest in knowing how many other users hold privileges to F at some point in time. (All these privileges ultimately originated of course in A.) A similar problem is the question through how many "hands" the right to F passed, until it arrived at some other user B. Both questions are fairly natural and of practical interest. The first question is concerned with the total number of users which at some point hold a certain right to F. Undoubtedly this is of considerable importance to a user of an authorization system. The second question asks for the length of the (shortest) chain of grants of privileges which enables B to hold this right to F. To see the practical importance of this question the reader should consider that most people are quite willing to trust a friend, they are probably less confident in trusting a friend's friend, and very few are willing to trust the friend of a friend of a friend; in other words, trust is not exactly a transitive relation! It should be pointed out that the two questions are not strongly related. To illustrate this, one might picture all the grants of a particular privilege to an object F at a given point in time as a tree, with the creator of F being the root and an edge going from A to B if A passed this right (to F) to B. Then the first question concerns the total number of nodes in the tree while the second question is related to the height of the tree. It is clear that the number of nodes in a tree is only marginally dependent on the height and vice versa. (Note that a node in this tree may have arbitrarily many children!)

It is not difficult to answer both questions in the given model (although it may be quite time consuming). However, given the original motivation for our questions, the user might also be interested in imposing *limits* which then are automatically enforced by the database management system. More specifically, suppose A creates F and passes some right r concerning F to B. When passing r, A also passes a nonnegative integer q, called the propagation number, which has the following significance: B can pass r on to some other user C only if $q > 0$, and furthermore the propagation number of any recipient C must be at least one less than the propagation number of the sender (for this right). This condition can easily be enforced by the system administrator. It follows that propagation number zero implies that the privilege can be *exercised* but not *granted*. This simple mechanism directly enforces that the length of the longest chain of grants of a particular right to a certain object can be effectively limited by the creator of this object.

We remark that implementing a limit on the total number of subjects which may hold a privilege to an object can be done analogously; the propagation number in this case would indicate the total number of subjects to which the recipient of the grant could pass on this privilege. There is, however, one difference, namely, at some point a subject may have exhausted its propagation number and is unable to make further grants. For example, if B received privilege *r* with propagation number 6 in this scheme and B granted *r* to C and D with propagation numbers 4 and 1, respectively, then B could not make any further grant of *r* as $6 = 4 + 1 + 1$ where the last $+1$ stems from the fact that B itself also holds the privilege. B could continue granting *r* only if it itself received an additional grant of *r*. Alternatively, B could "recover" privileges by revoking them from users to which B previously granted them. In the following we will concentrate exclusively on the first scheme, namely, a limit on the length of the chains of successive authorizations. To illustrate this concept, consider the following sequence of commands whereby we use the notation that R_6 signifies the right READ with propagation number 6 and R_* denotes READ with unbounded propagation number. (For example the creator of an object may have all rights on that object with unbounded propagation number.) Alternatively, there could be a system constant $M > 0$ which is the maximal permitted propagation number. Let us assume that A is the creator of the object F.

Time	Grantor	Grants			Grantee
1	A :	GRANT	R_{12}	ON F TO	B
2	B :	GRANT	R_{10}	ON F TO	D
20	A :	GRANT	R_2	ON F TO	C
21	C :	GRANT	R_1	ON F TO	D
22	D :	GRANT	R_6	ON F TO	E
23	E :	GRANT	R_2	ON F TO	X

It is clear that there is no need for specifying whether a grant is done with or without grant option, as this is governed in our extension by the value of the propagation number. The new scheme is not literally an extension of the old scheme as in the latter it was possible to have arbitrarily long chains, whereas in the former any chain is of a certain maximal length (unless R_*, for example, can legally be received—which in some sense would defeat the purpose of our modification in most cases).

It is easily verified that all of the commands are legal, i.e., each issuer has the required authorization and the conditions concerning the propagation of rights are satisfied, that is, no user tried to pass a right with a propagation number greater than or equal to its own. The resulting SYSAUTH table is then as follows (suitably modified to accommodate the extension):

USER	TABLE	TYPE	UPDATE	PRIVS	TIME
A	F	R	NONE	$R_*I_*D_*U_*X_*$	0
B	F	R	NONE	R_{12}	1
D	F	R	NONE	R_{10}	2
C	F	R	NONE	R_2	20
D	F	R	NONE	R_1	21
E	F	R	NONE	R_6	22
X	F	R	NONE	R_3	23

Recall that in the original authorization system as given by Griffith and Wade the revocation algorithm was not intuitively correct; for this reason it had to be modified as pointed out by Fagin. The extension introduced here requires yet another modification of the revocation procedure. To illustrate the salient points consider the following revocation command:

24 A : REVOKE R ON F FROM B

It follows that B loses R on F and therefore the grant from B to D is also to be revoked. Additionally, recall that D's grant to E is (partially) based on the now revoked grant from B to D. More specifically, since D now has R on F only with propagation number 1, it cannot grant R on F with propagation number 6. In other words, E's propagation number for R must be reduced to 0, and consequently E's grant of R to X must be revoked as E cannot grant a privilege it holds with propagation number zero. Thus the relational SYSAUTH will look as follows after execution of this revocation command:

USER	TABLE	TYPE	UPDATE	PRIVS	TIME
A	F	R	NONE	$R_*I_*D_*U_*X_*$	0
C	F	R	NONE	R_2	20
D	F	R	NONE	R_1	21
E	F	R	NONE	R_0	22

We leave it to the reader to formulate the appropriate revocation algorithm which will accommodate the extension as outlined.

BIBLIOGRAPHIC NOTE

The notion of bounded propagation of privileges was first introduced and discussed in Leiss.[55] A more detailed exposition of the underlying motivation as well as the required methods and algorithms can be found in Park.[68]

Exercises

1. Describe a recursive revocation algorithm (a) if the implementation is done using relational databases, (b) if the implementation is done using conventional databases. Estimate the time complexity of your algorithms. Compare these results with your results to Exercise 2, Section 3.4.3.

2. Consider the following sequence of commands whereby A is assumed to be the creator of object F:

1	A :	GRANT W_3	ON F TO	B
2	B :	GRANT W_2	ON F TO	C
3	B :	GRANT W_2	ON F TO	D
4	A :	GRANT W_4	ON F TO	C
5	C :	GRANT W_3	ON F TO	D
6	D :	GRANT W_a	ON F TO	E
7	B :	GRANT W_1	ON F TO	E
8	C :	GRANT W_1	ON F TO	E
9	A :	GRANT W_6	ON F TO	D
10	E :	GRANT W_b	ON F TO	X

Consider the following question:

(Q) Discuss which values for a and b ensure X can exercise W on F; which values for a and b ensure that X can grant W on F? (Clearly, only those values are to be considered which result in permitted commands.) For these values of a and b, determine the relation SYSAUTH!

(a) Answer question (Q).

(b) Assume that the next command is

11 A : REVOKE W ON F FROM D

Answer (Q) now.

(c) Assume as next command

12 A : REVOKE W ON F FROM C

Answer question (Q) now.

Chapter 4

Cryptosystems

4.1. Introduction and Motivation

4.1.1. The Notion of Encryption

Suppose two persons, A and B, wish to communicate, i.e., exchange messages. At first glance it appears that this is an easy enough task. Unfortunately, this appearance may be quite deceptive. The following problems arise:

(1) How can A and B make sure that nobody else can receive these messages?

(2) How can A (B) make sure that B (A) receives *all* of A's (all of B's) messages?

(3) How can A and B make sure that they receive only these messages and no others?

It should be clear that any of these three possibilities presents a real danger. As an illustration consider a chess game between three players where one plays simultaneously against the other two. Furthermore assume that the three players are in three different locations and that the moves are communicated in some way, e.g., by mail, telex, etc. Then problem (2) corresponds in our example to some move not being received by one of the players. This in itself would probably soon be detected as each of the two players involved would wait for the other one to make a move but it will certainly interrupt the normal course of a game. Problem (3) corresponds to a situation which may be much more difficult to detect, namely, an outsider (this may well be the third player!) might send one additional move to each of two players pretending to be the other legitimate player. In this case detection may take much longer. Considerably more complicated scenarios result from combining two or more subversions. An example of such a situation is an outsider intercepting one move [problem (2)] and replacing it with another move [problem (3)]. Again, it may take very long to detect this. This example also highlights an additional problem frequently encountered in this context: Detection of an abnormal situation does not imply that it can be determined who caused it. To make matters worse, one of the legitimate players, upon discovering at some point in the game that a certain move was wrong, might renege on it and pretend that an outsider was involved. Thus, to ensure orderly communication of confidential messages it is necessary that the authenticity of messages can be proved conclusively at some later point in time to any other party (e.g., an arbitrator or judge). Finally, to conclude our example, an outsider who wishes to monitor the game may be able to "tap" the communication line and get to know the moves [problem (1)] thereby gaining information about the strength of the players, their favorite strategies, etc.

Having convinced ourselves that there is in fact a multitude of problems with the notion of communication despite its deceptive simplicity and intuitive obviousness, and by extension having decided that it is desirable to resolve these problems, we now direct our attention to a possible solution. In what follows we will try to demonstrate that these three problems can be solved by using encryption methods, occasionally augmented by some suitable protocols.

First and foremost, encryption is a method of solving problem (1). Assume that A wants to send a message M to B and that A is afraid some eavesdropper X might be able to obtain the information contained in M. To prevent this, A modifies M in such a way that B is able to retrieve (the information in) the message M but X is not. Of all the methods for ensuring data security which are described in this book, this approach is easily the oldest one. Already in the Roman Empire, secret messages could be understood only by somebody who knew the code. All that had to be done was to make sure that the intended recipient of the message knew the code and that nobody else did. While the first codes were very simple compared to the present day codes, the basic idea is still the same: Modify the message in order to make it unintelligible to anyone but the intended recipient. No doubt, if we can achieve this, problem (1) is solved: We can prevent an outsider from receiving our messages; for while the eavesdropper may still be able to get access to the transmitted message, this will be of no use as it can be understood only by someone knowing the code, which by assumption is just the intended recipient.

Thus the main question now is how to modify a message in the way described above. What we need is a function E (for encrypt) which maps messages into encoded messages or ciphers. Without undue restriction we will assume that both message and cipher are over the same alphabet. This function E must have a number of properties; the most important one stipulates that it is possible to retrieve M when given $E(M)$, i.e., we must be able to decrypt the cipher. Consequently there must be another function D (for decrypt) such that

$$D(E(M)) = M, \qquad \text{for all messages } M$$

In the following we will only consider block ciphers. This implies that all messages to be encrypted are equally long and that E is applied to each of these blocks without taking encodings of previous blocks into account. This is in contrast to stream ciphers where each single symbol is encoded separately but the encryption depends on the encryption of the previous symbols, not only on the symbol to be encrypted. The main reason for this choice is that an error in a stream cipher will usually affect the whole message from this symbol on (thus usually rendering it incomprehensible), while in the block cipher only one block is affected but subsequent blocks of the same message are not. Clearly if a message is too long for one block it is split into several blocks of the specified size which then are transmitted in order. To ensure that certain blocks cannot be removed from such a blocked message by an eavesdropper, one can prefix each block with the note "This is block I of N" before the blocks are encoded.

It is important to note that E (and D) must be given in some form which allows effective computation given an argument, i.e., as an algorithm. Furthermore, given a message M, the computation of $C = E(M)$ as well as that of $M = D(C)$ should be possible within a reasonable amount of time, preferably in time linear in the length of the message. Naturally this excludes exceedingly complicated algorithms.

Given that many users want to encode not only one message but many and that the intended recipients of the messages are frequently different, it is unreasonable to expect that a new algorithm E is used for each of these messages or each of the recipients as this would be highly impractical. Consequently it makes more sense to have *one* algorithm E with many different *parameters*, called keys K, instead of many different algorithms. Thus the key K becomes another input, or argument, to the algorithm E. The same holds also for the decryption algorithm D. We reluctantly conclude that it is impossible to keep the algorithm E and D secret as a large number of persons may know them; instead we decide to keep the keys secret. Thus E and D are now functions of two arguments, keys in addition to messages and ciphers,

$$E(K, M) \qquad \text{and} \qquad D(L, C)$$

with K and L secret keys and E and D public algorithms. Following Simmons[90] we will distinguish symmetric encryption where L is essentially the same key as K, i.e., the same (or very similar) keys are used for encryption as well as for decryption, and asymmetric encryption where the two keys K and L are fundamentally different. (This will be explained in more detail below.)

Before we discuss further conditions on E and D we consider the possible attacks a cipher text will have to withstand in order to be considered unbreakable. We distinguish three kinds of attacks on a cipher.

(a) Ciphertext only attack: The attacker knows only the ciphertext C.

(b) Known plaintext attack: The attacker knows the ciphertext C and the corresponding plaintext M.

(c) Chosen plaintext attack: The attacker is able to submit any number of plaintext messages M (of the attacker's choice) and obtain their encodings C.

Any encryption method which cannot withstand a known plaintext attack is totally inadequate. Depending upon the circumstances not even this may be sufficient; possibly only an encryption method which resists a chosen plaintext attack may be acceptable.

In view of these observations the following are minimal requirements on any reasonable encryption method.

1. The probability of any symbol to appear in a ciphertext is equal to that of any other symbol, i.e., the probability is flat. If this requirement is violated the cryptanalyst can easily break the code by taking into account that different symbols have vastly different probabilities in the plaintext. For example, in a typical English text the letter e occurs about three times as often as the letters m or d and about ten times as often

as w or y or b. An extension of this argument leads immediately to the next requirement:

2. The probability of any n-tuple of symbols must be as flat as possible, at least for the first few values of n, $n = 2, 3, 4, \ldots$.

These two requirements have a rather startling implication. If a single bit of the plaintext is changed, the probability that any given bit of the resulting ciphertext is changed is about 1/2. Conversely a change of one bit in the ciphertext implies that the probability for any bit in the resulting decoded text to be changed is again about 1/2. In other words, a transmission error of just one bit in the ciphertext renders the decrypted message totally meaningless!

The present chapter is primarily concerned with secure encryption. By "secure" we mean that an outsider cannot break the code or cipher obtained by the encryption method. More precisely, an encryption method is secure if the key which is required for the decryption of the ciphertext cannot be determined by a known plaintext attack (possibly even by a chosen plaintext attack). Recall that we assume the algorithms E and D to be known to the attacker; surely if this is not the case our encryption methods can only become more but not less secure.

In the next two sections we will discuss methods of implementing functions E and D which satisfy the above requirements. However, a warning is in order: For all practical purposes, absolutely secure en- and decryption (i.e., a code which cannot be broken under any circumstances) is impossible. It is true that at least in theory absolute security can be attained; this is the method known as random one-time pad. Briefly an instance of this scheme can be stated as follows. Let M be a message represented in binary and let K be a true random sequence of zeros and ones of the same length as M. Furthermore each symbol of K must introduce as much uncertainty as is removed by a symbol of M (the key K must be incoherent). Then the corresponding cipher C is the exclusive or (bitwise) of M and K:

$$C = M \oplus K$$

In other words, if K has a zero in some position, C has the same bit as M in this position; if K has a one in a position, C has a one if M has a zero and a zero if M has a one in this position. There are two problems with random one-time pads: It is practically impossible to generate incoherent random sequences, and furthermore a method where the key is as long as the message and in which this key can only be used once is of little practical significance despite the fact that it can be proven to be unconditionally secure. Thus our keys will usually be substantially shorter than the message, and furthermore they are commonly used for en- and decrypting of more than one message. For these reasons we will have to settle for something less than absolute security, namely, we will call an encryption method secure if it is computationally impossible to break the code. By computationally impossible we mean the following: Suppose we want to determine the key involved in the decryption and assume we know that the fastest method for solving this problem requires about 2^k operations, where k is the length of the key K. Then we know that it is not impossible to break the code but for sufficiently large k it becomes computationally unfeasible. For example, if $k = 100$ then we have to perform more than 10^{30} operations; thus even if we could execute one hundred billion operations in one second we would still require more than three hundred billions years!

4.1.2. Applications

The most immediate aspect of en- and decryption is clearly that of secure communications [problem (1)], i.e., preventing unauthorized parties from "listening in" on confidential communication. In other words, encryption is a vehicle to render sensitive messages secret in spite of insecure communication channels. Another area of rapidly increasing importance is the protection of confidential data in a database. While it is theoretically possible to construct secure operating systems (see Chapter 3, Introduction) it is a fact of life that operating systems are notoriously unsafe. Thus it is relatively incautious to leave confidential data unprotected in a file which can be accessed via the operating system. In order to prevent this undesirable situation one can store encrypted data instead of plaintext data. The encryption can either be done on an intelligent terminal which is under the user's direct control, by a program which runs under the supervision of the (unsafe) operating system but which uses a user-supplied key for encrypting (thereby reducing the chance that an attacker with control over the operating system can get hold of the key), or a combination of both. If the emphasis is more on data integrity (no unauthorized user can alter the data) rather than on data security (no unauthorized user can access the data) encryption combined with redundancy in the data (additional check bits or storing the same data several times such that the blocks are not duplicated) can achieve detection or even correction of such illegal alterations. An interesting instance of this application is that of a confidential electronic mail system.

A third area where encryption is very useful is authentication and its more demanding version, digital signatures. The problem of authentication can be described as follows: If A and B want to communicate it is necessary that A can convince B that it is really A who is talking to B and vice versa. Any error here might result in B's releasing information to someone falsely pretending to be A. This is basically problem (3) mentioned above. The usual log-on procedure on a computer facility is an example of one-way authentication; while to the user it is quite obvious to whom s/he is talking, it is necessary that the computer knows exactly who is talking to it. On the other hand, to establish a connection between two nodes in a computer network is a problem of two-way authentication; both nodes must verify whether they are really talking to whomever they want to talk. Digital signatures are discussed later on; suffice it to say now that we can use encryption methods together with a suitable protocol for digital signatures. However, we want to point out why digital signatures are more than just simple authentication. Digital signatures must satisfy a number of requirements. First of all, they must act as authentication. However, this is not enough. A very important stipulation is that the signature must depend on the message to which it is appended. For suppose this were not the case. Then anybody could simply remove the signature from a genuinely signed message and append it to some faked message. If paper and pencil are involved this "cutting and pasting" will be discovered immediately; however, when messages are electronically transmitted this is not possible any longer. Finally, there is the issue of repudiation. Not only is it conceivable that the sender may not want to acknowledge the signature, it is also possible that the sender may deliberately lose the key(s) used to sign the message and then claim that somebody else forged the message and signed it with the lost key. Conversely, it is possible that B claims that A sent a signed message while in reality B falsified the message and A's signature.

A further application of encryption is related to the protection of proprietary software and data. Somewhat oversimplified the problem is the prevention of theft of such products. Typically, this danger occurs when software, e.g., compilers or payroll programs, is provided by their owner, who retains all proprietary rights to this product. Clearly the recipient will want to use the software and furthermore also be able to make certain modifications to it; thus the approach to treat the program as an immutable black box is not feasible in most instances (but consider "turn-key" systems). On the other hand, the proprietor is less than thrilled by the thought that the recipient may in turn sell the software to other interested parties (obviously without letting the owner partake in the profit). This problem has two facets, namely, prevention of such theft as well as its detection. In the first case it is made impossible (or at least very difficult) to resell the software; this is by far the most attractive approach. Unfortunately, it is also extremely difficult to realize. In the second case the idea is that the owner somehow learns that his/her product is used at an installation to which the owner did not give it; now the task is to *prove* that this program is in fact the owner's and not written independently by the accused. Below we will offer solutions to these problems. While there is no question that they can be implemented, many users may not be willing to pay the corresponding price, which is admittedly quite high.

4.1.3. Limitations

Below we discuss a number of limitations of the encryption methods. Most of them center on the application of encryption for the protection of data in a database system. One of the major drawbacks of storing encrypted data is that it is very difficult to perform operations such as comparisons, arithmetic, etc., on the encrypted data. In some cases privacy homomorphisms can be employed but their use is rather restricted (see Section 4.3.3). Most often the data can only be processed in its decrypted form, and this can obviously present problems as far as security is concerned. Another problem is connected with the use of redundancy in order to detect or even correct unauthorized modifications of the data. The difficulty here is that in very large databases where data items are relatively infrequently accessed it may take very long until these changes are detected. If the damage is great even error-correcting redundancies may not be sufficiently powerful to restore the data; the file will have to be retrieved in its entirety from old files. In order to limit the time interval during which the alterations remain unrecognized and in an effort to track down violators, extensive (and expensive!) updates may have to be performed periodically. The last problem we will mention here is key management. Revocation of keys, for instance, requires comprehensive decryption and subsequent reencryption with the new keys(s). Note that revocation will be necessary if someone loses a key. Clearly, losing a key in our context is considerably more troublesome than losing, say, the key to one's home; losing the encryption key requires reencryption with a new key, which in the analogy with the house key would correspond to changing every single item in one's house! Another difficult aspect of key management is that each item which is to be protected independently must have its own key. Thus it is possible that the number of keys is quite large. These keys must be stored somewhere; consequently the location of these keys must be extremely secure, strictly more secure than any of the files protected by these keys! (It must be strictly more secure since absolute security does not exist for all practical purposes, and an attacker who wants to get access to a particular file would otherwise be much better off in attacking the location of the keys, thereby obtaining the additional advantage that now *all* the files are accessible to him/her!) Furthermore in order to start communications, keys must be transmitted. In the case of symmetric encryption schemes this must be done via other (secure) channels, e.g., courier, in order to guarantee that the key(s) are not known to anybody else; in the case of asymmetric encryption schemes, there is a problem of authentication as far as the public keys are concerned. This will become clear later.

Finally it must be recognized that there are inherent limitations on our concept to come up with encryption schemes as secure as possible. Namely, we can state the following:

Theorem 29. The complexity of an encryption scheme whose en- and decryption can be done in P time (deterministic polynomial time) cannot be grater than NP time (nondeterministic polynomial time).

Proof. Simply guess the key nondeterministically and then verify (in polynomial time) that it is correct. □

We conclude this section with an overview of the remainder of this chapter. In Section 4.2 we outline the classical methods, namely, symmetric encryption schemes, that is, similar keys are used for encryption and decryption. It follows that these keys must be secret. Starting from the commercially most important symmetric encryption method, the Data Encryption Standard (DES), put forward by the United States National Bureau of Standards, we outline an improvement which has a much longer key and thus is considerably more secure.

Section 4.3 concentrates on public cryptosystems. These are asymmetric encryption schemes, thus there are two different keys, K and K', where K is used to encrypt,

$$C = E(M, K)$$

K' is used to decrypt,

$$M = D(C, K')$$

and K and K' are substantially different. The practically most important difference to symmetric encryption is that one of the keys may be public as long as the other one remains secret.

In Section 4.4 we describe ways of generating digital signatures. Both symmetric and asymmetric schemes can be used for this purpose. In both cases a certain protocol is necessary in order to guarantee that the requirements concerning digital signatures are satisfied.

Finally, in Section 4.5 we discuss methods which can be used in the attempt to protect proprietary software and data. Both detection of theft as well as prevention will be considered.

Exercises

1. Suppose an encryption method has the property that the probability for any combination of n symbols to occur in the ciphertext is flat. Prove that changing a single bit of the plain text (cipher text) implies a change in about half of all the bits of the cipher text (plain text).

2. Describe a scheme of introducing redundancies into stored encrypted data which will guarantee that subversion (a) is detected, (b) is corrected. Hint—Consider error-detecting and error-correcting codes. Treat the attempted subversion as errors. You will have to make assumptions about the subversion in order to solve this question; make sure to state all your assumptions explicitly!

Bibliographic Note

There are a number of good survey articles which either are entirely devoted to cryptography or cover it within the context of data security; most of them contain extensive bibliographies. We mention just three; further references can be found there. Denning and Denning[23] provide a survey of data security in general, Popek and Kline[70,71] are concerned with cryptography as it can be used for secure computer networks, and Simmons[89] gives an excellent overview of the state of cryptography today, contrasting classical methods (symmetric encryption) with the "new" methods (asymmetric encryption). Finally, Theorem 29 is taken from Diffie and Hellman,[26] the paper which started the new methods.

4.2. Symmetric Cryptosystems

In this section we will outline classical encryption methods, namely symmetric encryption. Recall that this means the use of the same (or similar) keys for encryption and for decryption. More specifically, if the keys are not the same then knowing one of them will enable us to determine the other with little effort. Since we agreed that the encryption and decryption algorithms cannot be assumed secret it is obvious that the keys must be kept secret. This is in contrast to the later-to-be-discussed asymmetric encryption methods where one of the two keys may in fact be public without endangering the safety of the encryption.

There are two primary operations involved in virtually all symmetric cryptosystems: substitution and transposition. Most acceptable symmetric encryption schemes employ both of them. In the operation substitution we have a bijection (one-to-one correspondence) mapping the alphabet onto itself. Recall that we agreed on having the same alphabet for the plaintext as well as for the ciphertext. A simple example is the function which maps (cyclicly) every letter of the alphabet to its subsequent letter (and Z to A); thus applying this encoding to

DATA SECURITY

yields

EBUB TFDVSJUZ

Various classes of such functions can be defined; the key then is used to select a particular function from such a class. Decryption is thus simply the inverse of this substitution function. It should be clear that this inverse is easily determined given the original encoding function.

While substitution is defined in terms of the letters of the underlying alphabet, the operation transposition is defined in terms of the positions of the given plaintext. In the simplest case the key is a permutation and the result of the operation is the application of this permutation to the plaintext. For example if we are given the permutation

$$\begin{matrix} 1 & 2 & 3 & 4 & 5 & 6 \\ 6 & 1 & 4 & 5 & 3 & 2 \end{matrix}$$

(key 614532) and apply it to DATA SECURITY we obtain

EDAS TAYCITRU

We now also see immediately how we can decrypt when given the key;

all we have to do is to compute the inverse of the permutation, which is a trivial operation, and apply this new permutation to the cipher. In the example the decryption key is

265341

It is clear that the operation substitution alone will not pass the test for a flat probability of the individual symbols of the underlying alphabet. In our example the letter E is mapped to the letter F and so will every other E. This observation also holds for the simple transpositions mentioned above, as can be easily verified. Consequently, combinating the simple substitutions and transpositions we discussed so far will not alleviate this problem.

An improvement are the simple Vigenère ciphers whereby the substitution rule (i.e., key) is a word (or a text) and the encryption process changes each letter depending on the letter in the same position in the given key. For example, if the chosen word is KEY and the plaintext DATA SECURITY, one would write the key repeatedly under the text letter for letter and use this string (which is now of the same length as the plaintext) to encrypt. In our example we get

plaintext	:	DATA SECURITY
key	:	KEYK EYKEYKEY
ciphertext	:	NERK WCMYPSXW

The ciphertext is obtained by adding the numbers of the letters in the same position for the plaintext and for the (extended) key together to get the number of the letter in the same position of the ciphertext (if necessary modulo the cardinality of the alphabet). In our case the cardinality of the alphabet is 26, the number of the letter D is 3 and the number of K is 10 (A is assumed to have number 0). Thus the first letter of the ciphertext has number 13 and is therefore N. Similarly, the last letter of the ciphertext has number 22 [namely, $(24 + 24) \bmod 26$] and thus is W. When sufficient care is applied in the choice of the key the cryptosecurity of the simple Vigenère ciphers can be quite good. In fact the one-time pads mentioned in Section 4.1.1 are an extreme case of this type of ciphers where the key is as long as the message and additionally satisfies other requirements. Decryption of simple Vigenère ciphers is again very easy if the encryption key is known, namely, it consists of subtracting (modulo the cardinality of the alphabet) the encryption key.

An even more powerful cipher is based on L.S. Hill's observation that linear transformations of the message space provide an elegant model of many cryptosystems. In this cipher a plaintext with n positions is represented as an n-tuple of integers (namely, the numbers of the symbols), encryption is a linear transformation, and decryption is its inverse. Clearly a minimum requirement is that the transformation be nonsingular. For example, let the letters A through Z be denoted by 0 through 25, the blank b̸ by 26, and the period and the comma by 27 and 28, respectively. Thus we have 29 different symbols. Since it is sufficient for us to know the number of a symbol we can perform all operations modulo 29; note that this is possible without complications since 29 is a prime number. Now consider the (2, 2) encryption matrix

$$\begin{bmatrix} 1 & 2 \\ 3 & 4 \end{bmatrix}$$

this matrix is nonsingular, its decryption matrix is

$$\begin{bmatrix} 27 & 1 \\ 16 & 14 \end{bmatrix}$$

(all operations modulo 29). If we want to encode the message

DATAb̸SECURITY.

we will split it into the following four blocks

DATA, b̸SEC, URIT, Y.b̸b̸

The corresponding (2, 2) matrices are therefore

$$\begin{bmatrix} 3 & 0 \\ 19 & 0 \end{bmatrix} \quad \begin{bmatrix} 26 & 18 \\ 4 & 2 \end{bmatrix} \quad \begin{bmatrix} 20 & 17 \\ 8 & 19 \end{bmatrix} \quad \begin{bmatrix} 24 & 27 \\ 26 & 26 \end{bmatrix}$$

Multiplying by the encryption matrix yields

$$\begin{bmatrix} 12 & 0 \\ 27 & 0 \end{bmatrix} \quad \begin{bmatrix} 5 & 22 \\ 7 & 4 \end{bmatrix} \quad \begin{bmatrix} 7 & 26 \\ 5 & 11 \end{bmatrix} \quad \begin{bmatrix} 18 & 21 \\ 2 & 11 \end{bmatrix}$$

or

MA.AFWEHb̸FLSVCL

It can be easily verified that given this text the decryption matrix will correctly produce the original plaintext.

The commercially important Data Encryption Standard (DES) proposed by the National Bureau of Standards is a well-known example of a symmetric cryptosystem. It is a scheme in which blocks of 64 bits are en- and decrypted; this is governed by a key of 56 bits which contains eight additional bits used for parity checks. A 64-bit block to be encoded is subjected to 16 applications of (nonlinear) substitutions alternating with permutations. The DES is object of a considerable controversy centering on Hellman's accusation that the National Security Agency (NSA) had chosen the key length of 56 in order to enable it to break the code with present technology. Consequently it appears desirable to construct cryptosystems with larger keys. In the following we adopt the approach taken by J. B. Kam and G. I. Davida,[47] who define a complete substitution–permutation network which has a number of attractive properties. In particular, complete substitution–permutation networks satisfy a condition which intuitively implies that breaking the code is difficult. These encryption methods have substantially larger keys than the DES.

A substitution–permutation network consists of a number of rows or stages, each of which consists of a fixed number of substitution boxes. The substitution boxes implement substitutions and the connections between the outputs of any stage to the inputs of the subsequent stage implement the permutations. Thus a substitution–permutation network has three parameters, namely,

- n: the number of input (output) bits (= size of the block to be encoded).
- k: the number of input (output) bits of each substitution box.
- m: the number of substitution–permutation stages.

Consequently n/k substitution boxes are needed at each stage, or mn/k altogether (assuming that k divides n). Let us denote by S_{ij} the jth substitution box at the ith stage, for $j = 0, \ldots, n/k - 1$ and $i = 1, \ldots, m$. Each substitution box S_{ij} is a logical circuit implementing a bijection $f(= f_{ij})$ which maps k inputs consisting of zeros and ones to k outputs consisting of zeros and ones:

$$f : \{0, 1\}^k \rightarrow \{0, 1\}^k \text{ bijectively}$$

As the n outputs of stage i form the n inputs of stage $i + 1$ for $i = 1, \ldots, m - 1$, it is clear that the substitution–permutation network itself is a bijection. Since by our general assumption the internal design of the network is known we will offer a choice which can be governed by a key; this is done by letting each S_{ij} consist of *two* substitution boxes, S'_{ij} and S''_{ij}. As there is a total of mn/k substitution boxes, the length of our key will be just mn/k. Each double box S_{ij} corresponds to one bit in the key and we choose S'_{ij} if its bit in the key is zero; if this bit is one we choose S''_{ij}.

Note that so far we have not made any assumptions about the S'_{ij} and S''_{ij}; at this point they could all be the identity. Thus we have to construct substitution–permutation networks which can withstand a known plaintext attack. In order to achieve this it is desirable to involve "as many input bits as possible" in the computation of each output bit (= bit of the ciphertext). To make this more precise, let $P = p_0, \ldots, p_{n-1}$ be the given plaintext and let $C = c_0, \ldots, c_{n-1}$ be the obtained ciphertext. Then we want to ensure that for every possible value of the key K, every output bit c_s depends on the values of all (!) the input bits $p_0, \ldots, p_{n-1}$ (cf. Exercises 1 and 2 on p. 168).

Formally, we call a substitution box (a substitution–permutation network) *complete* if the function represented by this box (by this network) is complete. Furthermore a bijective function f of n arguments is complete, if for all i and j in $\{0, \ldots, n-1\}$ there exist two n-bit vectors X_1 and X_2 such that X_1 and X_2 differ exactly in the ith bit, while $f(X_1)$ and $f(X_2)$ differ in at least the jth bit. It is not difficult to see that the notion of completeness of a substitution–permutation network captures the property informally stated above.

We now present an algorithm for constructing arbitrarily large complete substitution–permutation networks. We use the following notational convention: The input bits and the output bits of a single substitution box are numbered from 0 to $k-1$, and if we are at stage a of a substitution–permutation network, i.e., the boxes S_{aj} are involved for $j = 0, \ldots, n/k - 1$, then the ith input bit of S_{aj} is called the (a, j, i) bit of this stage and the ith output bit of S_{aj} is called the $(a, j, i)'$ bit. The algorithm is given in Figure 4.1.

It is clear that the result of the algorithm is a substitution–permutation network which encodes blocks of $n = k^s$ bits in s stages. It can furthermore be shown that this network is complete; for the proof we refer the reader to the original paper by Kam and Davida.[47]

However, we have not yet achieved much as far as the insufficient length of the key is concerned; recall that this was the original criticism directed against the Data Encryption Standard. For example, if the block size n is 1024 and the number k of input and output bits of each box is 32,

ALGORITHM

Purpose: Construct a complete n-bit substitution–permutation network using k-bit complete substitution boxes, with

$$n = k^s \qquad \text{for } s \geq 1 \text{ and } k > 3$$

Input: s stages of complete substitution boxes, the stages are numbered 1 through s, and each stage has precisely k^{s-1} substitution boxes. Furthermore it is assumed that each box consists of two boxes (see text).

Output: A complete n-bit substitution–permutation network.

Method: The following loop connects the inputs of the STAGEth stage to the outputs of the (STAGE-1)th stage.

```
for STAGE: = 2 step 1 until s do
    /* partition the boxes in the STAGEth stage into groups of k^(STAGE-1) boxes
    each                                                                     */
  for GROUP-OFFSET: = 0 step k^(STAGE-1) until k^(s-1) - k^(STAGE-1) do
      /* connect the input bits with number 0 of the k^(STAGE-1) substitution boxes
         in each group to the first k^(STAGE-1) output bits of the same group of the
         previous stage, then connect the input bits with number 1 of the k^(STAGE-1)
         substitution boxes to the next k^(STAGE-1) output bits of the group of the
         previous stage, etc.                                                */
    for BIT: = 0 step 1 until k - 1 do
        BIT-OFFSET: = BIT*k^(STAGE-1);
      for BOX: = 0 step 1 until k^(STAGE-1) do
          PRESENT-BOX: = GROUP-OFFSET + BOX;
          LAST-BOX: = GROUP-OFFSET + ⌊(BOX + BIT-OFFSET)/k⌋;
          LAST-BIT: = remainder(BOX/k);
          connect the input bit (STAGE, PRESENT-BOX, BIT) to the output
          bit (STAGE-1, LAST-BOX, LAST-BIT)'
      od
    od
  od
od
```

Fig. 4.1. Construction of Complete Substitution–Permutation Networks.

our key has merely 64 bits. Consequently we must provide a method for increasing the key size. This can be done as follows.

Suppose we have already constructed the following complete substitution–permutation network, using the algorithm given in Figure 4.1:

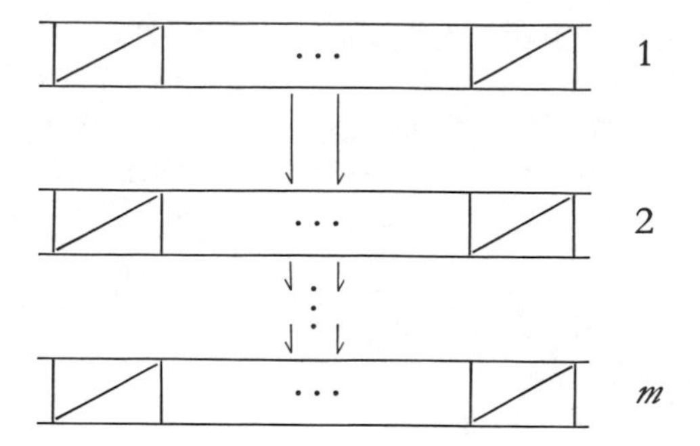

Then we combine the n output bits of each of the m stages and arbitrary n-bit vectors V_h by means of an exclusive or ($\oplus$):

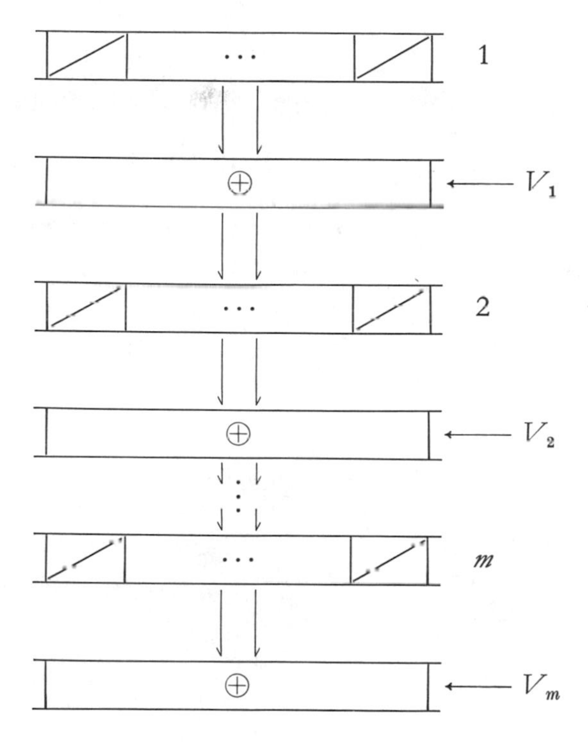

Each of the V_h is a key of size n, to be provided by the user. Thus the key size is substantially increased, namely, by a factor of $k + 1$. Furthermore, it is not difficult to see that the resulting network is complete provided the network we started with was complete. This follows directly from the observation that a complete function when exclusive-ored with any n-bit vector will again be complete.

The major advantage of this encryption scheme, which incidentally also applies to the DES, is that it is very fast. Thus it can be used to encrypt high-volume high-speed transmission of confidential data. This is of considerable importance especially as such an encryption scheme can easily be implemented on a chip thereby reducing the cost drastically.

Exercises

1. Consider the encryption matrix

$$\begin{bmatrix} 1 & 2 & 3 \\ 0 & 2 & 4 \\ 1 & 2 & 4 \end{bmatrix}$$

(a) Encrypt the message

ƀSYMMETRICƀƀCRYPTOSYSTEMS.ƀ

assuming the letters are numbered as described above.

(b) Compute the inverse (modulo 29) of this matrix. Remember that the entries of the inverse matrix are all integers between 0 and 28.

(c) Decrypt the following ciphertext:

Oƀ.WAOLOCQWFIƀXXBX

2. Show that the simple transpositions as discussed in the text can be modeled by Hill's linear transformations. Hint–Given the permutation (614532), the corresponding Hill transformation is given by

$$\begin{bmatrix} 0 & 0 & 0 & 0 & 0 & 1 \\ 1 & 0 & 0 & 0 & 0 & 0 \\ 0 & 0 & 0 & 1 & 0 & 0 \\ 0 & 0 & 0 & 0 & 1 & 0 \\ 0 & 0 & 1 & 0 & 0 & 0 \\ 0 & 1 & 0 & 0 & 0 & 0 \end{bmatrix}$$

Describe precisely how to en- and decrypt!

3. Prove by induction on the number of stages that the algorithm in Figure 4.1 produces a complete function.

4. Let g be a complete function,

$$g : \{0, 1\}^n \to \{0, 1\}^n$$

and let v be an element of $\{0, 1\}^n$. Prove that

$$g(x) \oplus v$$

is again a complete function of x.

5. Prove that two simple substitutions can be replaced by a single substitution.

6. Prove that two linear nonsingular transformations can be replaced by a single nonsingular transformation.

BIBLIOGRAPHIC NOTE

For more information about symmetric encryption schemes including additional references see Simmons.(89) The DES was announced in Roberts;(81) the subsequently ensuing controversy is reflected in Diffie and Hellman,(27) Morris *et al.*,(66) and Sugarman.(94) The notion of a complete substitution–permutation network and the constructions given in this section are from Kam and Davida.(47)

4.3. Public Key Cryptosystems

In this section we discuss asymmetric encryption methods, often called public key cryptosystems. These cryptosystems derive their name from the fact that of the two keys which are involved in encryption and in decryption, one may be publicly known while the other one must be kept secret. This is in contrast to symmetric encryption methods where *both* keys must remain secret. We first give an overview of the concept of a public key cryptosystem which is due to Diffie and Hellman. Then we present the basic idea of an implementation due to Rivest, Shamir, and Adleman. This implementation is based on the observation that factoring a large integer is apparently a considerably more difficult operation than testing the same number for primality. This relation will be made more precise using a variant of this scheme due to M.O. Rabin. Then we outline another implementation due to Merkle and Hellman which is based on the knapsack problem. This problem is known to be NP-complete in general. We conclude this section by indicating to what extent some encryption methods permit operations on encrypted data; this is the idea of privacy homomorphisms proposed by Rivest, Adleman, and Dertouzos.

4.3.1. The Concept of Public Key Cryptography

As in the case of symmetric encryption we will assume that the functions (algorithms) E and D are known. However, while before we had a single key K for encryption and decryption (or one key could easily be computed given the other), now we will assume that the two keys K (for encryption) and L (for decryption) are "substantially different". By this we mean that given one of the two keys K and L, it is computationally impossible to determine the other key. In this section we will only outline the concept; concrete implementations are postponed until the next sections.

In a public key cryptosystem all users employ the same encryption algorithm E and the same decryption algorithm D. Each user A possesses a pair of keys (K_A, L_A) where K_A is A's encryption key and L_A is A's decryption key, i.e.,

$$D(E(M, K_A), L_A) = M \qquad \text{for all messages } M$$

In fact, in many cases it is convenient to require that the two keys are interchangeable, i.e.,

$$E(D(M, L_A), K_A) = D(E(M, K_A), L_A) = M$$

The important new idea (due to Diffie and Hellman) is that each user A places his/her encryption key K_A in a public file while keeping the decryption key L_A secret. Thus every user knows anybody's encryption key but only his/her own decryption key. This implies that a user A must communicate the key K_A in a secure way to the public file administrator; otherwise problems of authentication will arise from the onset.

Suppose now that some user, say, B, wants to send a message M to A. B will look up A's encryption key K_A and send the cipher

$$C = E(M, K_A)$$

to A; this is possible for K_A is public. Now A applies his/her secret decryption key to the received cipher C,

$$D(C, L_A) = M$$

and thus retrieves M. We claim that this solves problem (1) mentioned in Section 4.1. Note that C is transmitted; C is encrypted with A's key K_A. In order to retrieve M from C one must decrypt C. However, to do this, A's decryption key L_A is required which by assumption is known only to A. This shows that only A, the intended recipient, can "read" the message; eavesdropping is not possible. An analogous scheme is employed if A wants to send a message to B.

The most important implications of this scheme are as follows:

1. If two users wish to communicate with each other it is not necessary (as it is in symmetric encryption schemes) that there be an initial exchange of keys between them via a secure channel.

2. Only two keys are required per user, independent of the number of participants in such a system. In the conventional schemes, if n users want to communicate independently and securely with each other, a maximum of $n(n-1)/2$ keys is required.

The public file which contains all the encryption keys will require considerable protection; the by far most crucial property to be attained is that of data integrity, i.e., keys and the names of the users associated with them must not be updatable by any unauthorized user. This public file will play the same role for secure communications as the telephone directory for spoken conversation; it will hold names and the corresponding keys necessary to establish communication.

We only mention here that with such a public key cryptosystem the problem of authentication and even that of digital signatures can be efficiently solved; a discussion of these issues is postponed until later.

4.3.2. An Implementation Based on Factoring and Primes

In this section a concrete implementation of the concept presented above will be discussed. Before we outline the method, a few words about the requirements are in order, especially as far as complexity considerations are concerned. As already pointed out, in view of the fact that absolute security is practically impossible to achieve we settle for computational security (i.e., it is computationally impossible to break the code). In addition to the already stated requirements (speed of en-

and decryption, flat distributions of the probabilities of the symbols) we now must also insist that determining the only secret piece in public key cryptosystems, the decryption key L, is computationally impossible given the encryption algorithm E, the decryption algorithm D, and the (public) encryption key K. Furthermore, if this method is to be of any practical value it must be easy to compute pairs of keys (K_A, L_A) for an arbitrary number of users A.

The method outlined below satisfies these requirements. The central observation is the apparent discrepancy between the time complexities of two seemingly closely related operations, namely, testing whether a given number is a prime and factoring this number. Indeed, the best algorithm presently known will determine whether a given l-digit number is a prime in time polynomial in l, while for factoring there is no known polynomial-time algorithm. More specifically, one can find an l-digit prime in time $O(l^4)$ if one is willing to assume the extended Riemann hypothesis, and such a prime can be found even faster if one is willing to accept a probabilistic algorithm for testing primality, i.e., the answer is always correct if it is "not prime" and it is correct with arbitrary probability (that is, with a probability of the user's choosing) if it is "prime." [We remark that without these assumptions, namely, extended Riemann hypothesis or probabilistic methods, it is still not known whether primality can be tested in polynomial time (see, for instance, Adleman[1]). However, this is not a very important issue here, as for our purposes these fast methods (and the answers they produce) are quite acceptable.] To put these observations about the complexities of the two problems into proper perspective, testing whether a given 200-digit number if prime may require about one minute, but factoring this very same number (assuming that one operation requires ten nanoseconds) will take several million years!

The encryption method based on factoring and primality is as follows. One chooses a pair of l-digit primes, p and q, with l so large that factoring the number n is computationally infeasible where n is the product of p and q ($n = pq$). Furthermore one chooses two numbers e and d such that the following holds: (i) e and $(p - 1)(q - 1)$ are relatively prime (have no common divisor greater than 1), and (ii) $ed = 1 \bmod ((p - 1)(q - 1))$. Note that $(p - 1)(q - 1)$ is the Euler quotient of n, that is, the number of integers less than n and relatively prime to n. The set of residue classes modulo $(p - 1)(q - 1)$ forms a group under multiplication; consequently the second requirement states that e and d are multiplicative inverses. In order to encrypt a message M one raises

M to the power of e modulo n:

$$C = M^e \bmod n$$

Decryption is then achieved by raising the ciphertext C to the power of d, the multiplicative inverse of e, modulo n:

$$M = C^d \bmod n$$

This implies that the keys involved are e and d, e for encryption and d for decryption, in addition to the number n. Furthermore we see that n must have no fewer digits than the message M to be encoded, or conversely, n must be chosen in such a way that it is larger than the largest number representable by the message (M treated as one binary number). This is not unreasonable as one will use blocking anyway. Consequently a user of such a public key cryptosystem will publish n and e and will keep d secret. This enables every user to encrypt but nobody who does not know d can decrypt. Thus a user who wishes to participate in a public key cryptosystem must first generate two l-digit prime numbers; this can be done in at most $O(l^3)$ operations. Then e and d must be determined. For e we start with a randomly chosen number and test whether the greatest common divisor $\mathrm{GCD}(e, (p - 1)(q - 1))$ is 1, using Euclid's algorithm. If it is we are finished, if not we can for instance increase e by 1 and iterate the method. After a small number of iterations we will obtain a value for e which is relatively prime to $(p - 1)(q - 1)$. Also note that Euclid's algorithm is very efficient. To determine d such that $ed \equiv 1 \bmod ((p - 1)(q - 1))$ we apply Euclid's algorithm once more. Thus the complexity of finding the required pairs of keys is certainly of polynomial time in the length of the number n ($= pq$).

The next issue of importance is the complexity of the operations encrypt and decrypt. We note that raising a number x to the eth power modulo n requires no more than

$$O(\log(n)T(n))$$

operations for $e < n$, where $T(n)$ denotes the time complexity required to multiply two numbers modulo n. This bound comes from an obvious extension of the observation that the operation of raising to the power of say, 8, which involves 7 multiplications, can be replaced by three multiplications, namely, squaring the number, then squaring the result, and squaring this result again. Thus while $O(\log(n)T(n))$ is not linear in the length of n, encryption can still be considered an efficient algorithm. The same holds of course for decryption too.

Example. Let $p = 17$ and $q = 19$, thus $n = 323$ and $(p - 1)(q - 1) = 288$. We choose for the encryption exponent e the value 5; this clearly satisfies

$$\mathrm{GCD}(e, (p-1)(q-1)) = 1$$

We now have to determine the decryption exponent d such that

$$5*d = 1 \mathrm{mod}(288)$$

Using Euclid's algorithm one finds

$$d = 173$$

(one easily verifies directly that $ed = 865 = 3*288 + 1$). More specifically, this is done as follows.

Euclid's algorithm computes $\mathrm{GCD}(x_0, x_1)$ for $x_0 \geq x_1$ by computing a series of integers $x_0, x_1, x_2, \ldots, x_k$ for some k such that

$$x_0 > x_1 > x_2 > \cdots > x_k \geq 1$$
$$x_{i+1} = x_{i-1} \mathrm{mod}(x_i) \qquad \text{for } i = 1, \ldots, k$$
$$x_k = \mathrm{GCD}(x_0, x_1)$$

This algorithm can be modified to compute additionally for each x_i integers a_i and b_i (not necessarily positive) such that

$$x_i = a_i x_0 + b_i x_1$$

If $x_k = 1$ then b_k is the inverse of $x_0 \mathrm{mod} x_1$. In our example we have to determine GCD(288, 5). This yields

$x_0 = 288$	$a_0 = 1$	$b_0 = 0$	$x_0 = 1x_0 + 0x_1$
$x_1 = 5$	$a_1 = 0$	$b_1 = 1$	$x_1 = 0x_0 + 1x_1$
$x_2 = 3$	$a_2 = 1$	$b_2 = -57$	$x_2 = 1x_0 + (-57)x_1$
$x_3 = 2$	$a_3 = -1$	$b_3 = 58$	$x_3 = (-1)x_0 + 58x_1$
$x_4 = 1$	$a_4 = 2$	$b_4 = -115$	$x_4 = 2x_0 + (-115)x_1$

As $b_4 = -115 = 173 \mathrm{mod}(288)$ it follows that 173 is indeed the desired integer d between 0 and 287.

Suppose now that the message we want to send is

$$M = 256$$

We have to compute 265^5. This requires three multiplications modulo 323; first we make a list of powers of two of powers of 256:

$$256^1 = 256 \bmod(323)$$
$$256^2 = 290 \bmod(323)$$
$$256^4 = 290 \cdot 290 \bmod(323) = 120 \bmod(323)$$

Then, as 5 is 101 in binary we compute

$$256 \cdot 120 \bmod(323) = 35 \bmod(323)$$

Consequently the encrypted message to be sent is

$$C = 35$$

To decode, we again first make a list of powers of two of powers of 35:

$$35^1 = 35 \bmod(323)$$
$$35^2 = 256 \bmod(323)$$
$$35^4 = 290 \bmod(323)$$
$$35^8 = 120 \bmod(323)$$
$$35^{16} = 188 \bmod(323)$$
$$35^{32} = 137 \bmod(323)$$
$$36^{64} = 35 \bmod(323)$$
$$35^{128} = 256 \bmod(323)$$

As 173 is 10101101 in binary, we then compute

$$256 \cdot 173 \cdot 120 \cdot 290 \cdot 35 \bmod(323)$$

it is easily verified that this is, in fact, the message originally transmitted, namely, 256. Thus for decoding we required 11 multiplications modulo 323.

Finally we must pose the question whether this code can be broken, or more precisely whether it is computationally possible to break it or not. We observe that knowledge of p (or q) allows to determine d as e is known; simply divide n by p, which gives q, and then use Euclid's algorithm to determine a multiplicative inverse of e in the group of residue classes modulo $(p-1)(q-1)$. A number of restrictions are discussed in the literature which should be imposed on the primes p and q in order to ensure that factoring n is difficult (see Herlestam,[41] Simmons and Norris,[90] and Williams and Schmid[99]); it appears that it is not difficult to satisfy them. However, so far there is no proof that decryption in this scheme is as hard as factorization. A first step in this direction are the results by Rabin[73] and Williams[98] which show for a variant of the Rivest–Shamir–Adleman scheme that decryption is equivalent to factoring, assuming that the cipher is attacked by a ciphertext only attack. We will comment on this in the next section. It should be pointed out, however, that even a general proof of this statement (decryption without knowing d is as hard as factoring n) would not be completely satisfactory as we still do not know whether factoring is indeed exponential in the length of the number to be factored. While no better algorithms are known, this has not been proven to be a lower bound on the complexity either. Consequently it is conceivable that some day a method will be discovered which factors in polynomial time. Until that day, however, this code is unbreakable for all practical purposes provided we choose n of suitable length. At present, a length of 400, i.e., both p and q of length approximately 200, appears very satisfactory. As pointed out factoring such a number will take much more time than any of us will ever have.

4.3.3. Code Breaking and Factorization

The encryption method described in the last section is at most as difficult as factoring, since anybody who manages to factor n (into pq) will also be able (with negligible effort) to determine the private key d by applying the computations given in the previous section. However, it is conceivable that other decoding methods exist which do not depend on factorization and also do not imply factorization of n, although nobody so far has managed to come up with one. Thus it is possible that this method is strictly less difficult than factoring n.

Rabin discovered a variant of the Rivest–Shamir–Adleman scheme for which he was able to prove that the ability of decoding messages without knowing the (private) decoding key implies the ability to factor n with the same amount of computational effort. Since the other direction is clear (factoring implies decoding) this shows that for his method, factoring is *provably equivalent* to decoding. Before we proceed with his scheme we reiterate that this in itself does not imply that decoding is extremely difficult as the complexity of factoring is not known; there is, however, substantial evidence to the effect that factoring is intractable, i.e., not solvable in polynomial time in the number of digits of the integer to be factored.

An additional remark is in order about the use of complexity arguments in connection with cryptography. (For a somewhat differing assessment of this issue see Brassard[6-8]). The central issue in complexity is the worst-case behavior of some algorithm. This is not appropriate for our purposes; in fact not even the average-case complexity may be acceptable. To see this consider as an extreme and admittedly hypothetical example that for 50% of all keys an encryption function is extremely easy to break, say, in linear time in the length of the key, while the remaining 50% of the keys require exponential effort. Thus the worst case as well as the average case of breaking this code would be exponential. However, since about one half of all the encoding keys are completely inacceptable, this method would have to be rejected. Consequently, for our purposes the best-case complexity is more appropriate—although this approach may also be misleading, as one can see from the (again extreme) example where *one* of all the keys requires linear effort for breaking this encryption function while all other keys require exponential effort. Thus in addition to the conventional complexity arguments, one must also consider the fraction of keys for which a particular encryption function is easy to break. For the Rivest–Shamir–Adleman scheme a number of requirements are discussed in the literature which exclude these "easy" encoding functions.

In this connection we report Rivest's observation that Rabin's method (to be outlined shortly) is secure only when confronted with a known plaintext attack but *not* with a chosen plaintext attack.

While the Rivest–Shamir–Adleman scheme has a single decoding function D corresponding to each encoding function E, Rabin's scheme differs slightly in that there are *four* candidates for D given E. In other words, E is not quite a bijection. The encryption function E is defined as follows:

$$E(M) = M(M + b)\mathrm{mod}(n)$$

where as before n is the product of two approximately equally large primes, M is the message to be encoded, and b is an integer satisfying

$$0 \leq b < n$$

The public keys in this scheme are n and b, while p and q are private. The decoding function D can be obtained as follows. Given any message C, we want to determine M such that

$$C = M(M + b)\mathrm{mod}(n)$$

In fact it turns out that this congruence has four solutions for M; thus we want to find M_i for $i = 1, 2, 3, 4$. We first show that it suffices to solve the two congruences

$$C = M(M + b)\mathrm{mod}(p) \qquad \text{and} \qquad C = M(M + b)\mathrm{mod}(q)$$

to determine a solution to the original congruence. (This is of course possible as the legitimate decoder knows p and q!) Assume that a is an integer satisfying the congruences

$$a = 1\mathrm{mod}(p) \qquad \text{and} \qquad a = 0\mathrm{mod}(q)$$

Similarly, let b be an integer such that

$$b = 0\mathrm{mod}(p) \qquad \text{and} \qquad b = 1\mathrm{mod}(q)$$

Then one can verify that

$$z = ar + bs$$

is a solution of the original congruence (modulo n) if r is a solution of the congruence modulo p and s is a solution of the congruence modulo q. Thus one must solve congruences

$$x^2 + bx - c = 0\mathrm{mod}(p)$$

Such a congruence can be solved provided we are able to extract square roots modulo a prime p: Given any integer f, let sqrt(f) denote an integer g between 0 and $p - 1$ such that

$$g^2 = f\ \mathrm{mod}(p)$$

Note that there are two such integers satisfying this congruence; the other one is $p - g$. As p is a prime the square root always exists. Similarly, f/h is that integer g (as always in this section between 0 and $p - 1$) such that

$$gh = f\ \mathrm{mod}(p)$$

The integer $1/h$ is easily determined by Euclid's algorithm. It then follows that

$$-b/2 + \mathrm{sqrt}(c + b^2/4) \quad \text{and} \quad -b/2 - \mathrm{sqrt}(c + b^2/4)$$

are solutions of the congruence

$$x^2 + bx - c = 0 \mathrm{mod}(p)$$

Using an algorithm in Berlekamp[4] one can show that no more than $0(\log(p))$ operations modulo p are required to extract square roots modulo p. Thus it follows that decoding can be achieved in

$$0(\log(n))$$

operations, provided one knows p and q. Encoding of course requires only constant time.

Rabin then shows that breaking this encoding scheme is at least as hard as factoring n. More specifically, if there is an algorithm which solves the congruence

$$x^2 = m \mathrm{mod}(n)$$

in $F(n)$ operations then there exists an algorithm which factors n in

$$2(F(n) + \log(n))$$

operations. In fact, this result can be strengthened to include a statement about the fraction of keys for which the code is breakable, for Rabin also shows that given any algorithm which solves the congruence

$$x^2 = m \mathrm{mod}(n)$$

in $F(n)$ operations for a fraction of $1/h$ of all the numbers m between 0 and n for which their greatest common divisor with n is 1, this algorithm can be converted into an algorithm which factors n and requires

$$2(hF(n) + \log(n))$$

operations. Thus for Rabin's scheme it follows that factoring and breaking the code are computationally equivalent (equally hard) operations.

4.3.4. An Implementation Based on the Knapsack Problem

In this section we give an outline of a second implementation of public-key cryptosystems. This implementation is due to Merkle and

Hellman and employs the knapsack problem to generate ciphers which are difficult to break. We first explain this problem and then present the basic idea of their method.

Given $n + 1$ positive integers $a_1, \ldots, a_n, a$ the knapsack problem consists of selecting a subset of the $a_1, \ldots, a_n$ whose sum is precisely a. In other words, we have to find $x_1, \ldots, x_n$ with x_i in $\{0, 1\}$ such that

$$a_1x_1 + \cdots a_nx_n = a$$

It is known that determining whether a given knapsack problem (A, a) with A being the vector $(a_1, \ldots, a_n)$ has a solution is an NP-complete problem; thus it belongs to a class of problems for which no polynomial time algorithm is known, and furthermore existence of such an algorithm for any one of these problems would imply the existence of such an algorithm for all of them. However, the complexity of the knapsack problem depends greatly on the choice of A. As a simple example assume that A satisfies

$$a_i > a_1 + \cdots + a_{i-1} \qquad \text{for all } i = 2, \ldots, n$$

In this case there exists a simple algorithm to decide in linear time whether there is a solution, and if there is one then the algorithm will determine it.

The knapsack problem forms the basis of an implementation (realization) of the abstract notion of public cryptosystems. The idea is as follows. One starts with two large integers m and w with the property that they are relatively prime, i.e.,

$$\text{GCD}(w, m) = 1$$

Then one selects a knapsack vector which can be easily solved, for example, if

$$a_i > a_1 + \cdots a_{i-1} \qquad \text{for all } i = 2, \ldots, n$$

This vector is then transformed into a *trapdoor* knapsack vector B by letting

$$b_i = wa_i \bmod(m) \qquad \text{for } i = 1, \ldots, n$$

Consequently, knowledge of w and m allows easy computation of

$$a = (b/w)\bmod(m)$$

where (B, b) is assumed to be the given knapsack problem. Note that the quotient b/w is again an integer between 0 and $m - 1$ whose existence is guaranteed by the condition $\text{GCD}(w, m) = 1$. It can be easily determined by Euclid's algorithm. Since multiplication by the integer $1/w$ transforms B into A, we thus can reduce the given (difficult) problem (B, b) to the easy knapsack problem (A, a). Usually for technical reasons one also assumes that

$$m > a_1 + \cdots + a_n$$

Since the given knapsack problem (B, b) does not have as nice a structure as (A, a) it appears that the original problem is the more difficult of the two.

Before describing how this scheme is used for obtaining an encryption scheme, we outline a technique to further enhance security.† The idea is to iterate the operation of constructing a trapdoor knapsack. Recall that in the original scheme we passed from an easily solved knapsack problem (with vector A) to a difficult one by multiplying (modulo m) by an integer w [which satisfied $\text{GCD}(m, w) = 1$]. Obviously this process can be repeated, i.e., we may construct another knapsack vector C by defining

$$c_i = w'{}^*b_i \bmod(m') \qquad \text{for } i = 1, \ldots, n$$

where w' and m' are again large integers which are relatively prime. Obviously, given (C, c), w', and m', one can easily compute (B, b) and from this using w and m one can obtain the easy knapsack problem (A, a). There is no question that this process can be repeated as often as desired.

Intuitively it is quite clear that with each iteration the structure of the original problem (A, a) becomes more obscured. Unfortunately, intuition is frequently misleading. For example, applying two simple substitutions or two linear transformations (nonsingular) in sequence does not improve the security of the respective ciphers as in both cases two or more applications can be replaced by a single application. However, in the case at hand it turns out that our intuition is on the right track. This is because it is not possible in general to replace two iterations of this process by a single one. More formally, if we transform (A, a) into (B, b) using (w, m) and then transform (B, b) into (C, c) using (w', m'), then in general there do not exist w'' and m'' which will allow us to transform (A, a) into (C, c) in a single step. Consequently, any

† See also the remark at the end of this section.

body who wants to solve the problem (C, c) without knowing (w, m) and (w', m') is forced to solve two separate knapsack problems. Also, while it is usually not too difficult to verify that one obtained the correct plaintext ("Does it make sense?"), this definitely does not apply to the intermediate problem! Thus there is considerable evidence that iteration will greatly increase the difficulty of breaking a code based on this method.

We now come to the description of how the method can be used to encrypt. One starts with a trivially solvable knapsack vector A, e.g., one which satisfies

$$a_i > a_1 + \cdots + a_{i-1} \qquad \text{for } i = 2, \ldots, n$$

(If this condition is met we will call the a_i strictly dominant.) Then one chooses a pair (w, m) of large, relatively prime, integers. The assumption $\mathrm{GCD}(w, m)$ guarantees existence of another integer $1/w$ between 0 and $b - 1$ such that

$$w \cdot (1/w) = 1 \bmod (m)$$

The trapdoor knapsack vector B is computed by setting

$$b_i = wa_i \bmod (m) \qquad \text{for } i = 1, \ldots, n$$

This vector B will be the public key. The private keys are m and w. A message M is transmitted as follows. Assume M is encoded in binary and of length n, i.e.,

$$M = (m_1, \ldots, m_n) \qquad \text{for } m_i \text{ in } \{0, 1\}$$

Then the sender computes

$$b = m_1 b_1 + \cdots + m_n b_n$$

and transmits b as the encrypted message (ciphertext). Since B is public, anybody can *send* a message. However, decoding this message is easy only if one knows (w, m) since this allows us to transform B into A and then solve the easy knapsack problem $(A, b/w)$. An unauthorized receiver, on the other hand, would have to solve the given knapsack problem (B, b), which is computationally more difficult. This is especially true if one uses not only one transformation but two or more, as pointed out above. Note that it is not clear at all to the public whether one or more iterations are used as only the result (B) is published, but several iterations may be required to obtain it. All the pairs (w, m) are secret and therefore no obvious inference can be made as to the number of iterations.

Let us give a simple example. We start with the vector

$$A = (1 \ \ 3 \ \ 5 \ \ 11 \ \ 21)$$

and we choose $w = 37$ and $m = 43$. Consequently,

$$B = (37 \ \ 25 \ \ 13 \ \ 20 \ \ 3)$$

Furthermore, $1/w$ is 7. B is now published and (w, m) kept secret. Assume that we receive the following message:

$$14 \ \ 13 \ \ 20 \ \ 7 \ \ 40 \ \ 25 \ \ 17 \ \ 19$$

In order to decode it we first multiply (modulo 43) these numbers by 7 $(= 1/w)$, which produces

$$12 \ \ 5 \ \ 11 \ \ 6 \ \ 22 \ \ 3 \ \ 33 \ \ 4$$

Now we can solve the original (easy) knapsack problem (A, a) with a ranging over the numbers 12, . . . , 4. This gives the following binary vectors:

10010 00100 00010 10100 10001 01000 10011 11000

and if we denote A by 00000, B by 00001, . . . , and Z by 11001, then the message reads SECURITY.

Two remarks are in order. It is true that for a given knapsack problem (X, x) there may be more than one solution just as there may not be any solution. The fact that we are performing modulo-arithmetic may aggravate this problem. For example, with the vector A in the above example but with $w = 3$ and $m = 32$ one would get the message

$$4, \ 15, \ 1, \ 18, \ 2, \ 9, \ 3, \ 12$$

which would be decoded to SECURIQY. This is due to the fact that the last two components of A add up to m, and since the letter T corresponds to 10011 and Q corresponds to 10000 and these two vectors are equivalent modulo 32 as solutions, Q and T cannot be distinguished. The other remark concerns the size of the vector as well as its components. Clearly the numbers used in the example are far too small. Merkle and Hellman suggest that n be at least 100, that m be uniformly chosen from $\{2^{201}+1, \ldots, 2^{202}-1\}$, that the a_i be uniformly chosen from $\{1, \ldots, 2^{99} * 2^{100}\}$ (strictly dominant), and that w be chosen as the result of dividing a uniformly chosen integer v from $\{2, \ldots, m-2\}$ by GCD(v, m). Clearly, if higher security is required, increasing the size of the numbers is one obvious way to attain it.

Remark (Added in Proof)

In a draft, privately circulated in Spring 1982, Adi Shamir outlines a method which breaks the Merkle–Hellman implementation of public-key cryptography. More specifically, this paper shows the following: If one starts with a strictly dominant vector A and transforms it into the vector B by multiplying A by w mod(m), *almost all* ciphertexts b can be broken in polynomial time, i.e., for almost all given b such that $b = m_1b_1 + \cdots + m_nb_n$ the plaintext $m_1 \cdot \cdots \cdot m_n$ can be effectively determined. However, the method employed by Shamir in compromising the Merkle–Hellman scheme applies only to a single-step trap-door knapsack; in particular, it relies heavily on A's property to be strictly dominant. It does not seem to be possible to extend Shamir's method of compromise to a multi-step trap-door knapsack, i.e., a knapsack problem which starts with a strictly dominant knapsack vector $B^{(0)} := A$, but then transforms A into $B := B^{(1)}$ in l steps where $B^{(i)}$ is obtained from $B^{(i-1)}$ by multiplying by $w^{(i)}$ mod($m^{(i)}$) for suitably chosen integers $w^{(i)}$ and $m^{(i)}$, $i = 1, \ldots, l$, where $l \geq 2$. Note that a multi-step knapsack problem *cannot* be replaced by a single-step knapsack problem; see Exercise 6 at the end of this section. Consequently, Shamir's method does not state anything about the security of the multi-step knapsack implementation which is therefore still open.

4.3.5. Privacy Homomorphisms

In Section 4.1.3 we mentioned some limitations applicable to all encryption methods. A rather serious one is related to the use of encryption to protect data files from unauthorized access, another aspect of database security. Instead of just storing confidential data in a file system which would be subject to subversion of its data security coming from a variety of sources, most notably from an unsafe operating system which can be abused by malicious users, the suggestion is to store data only in encrypted form. A typical example of a situation where this might be desirable is that of a small savings and loan association which rents computing time from a time sharing company. As long as the appropriate key remains secret (for symmetric encryption the single key used for en- and decrypting, for asymmetric encryption just the decrypt key) there will be no problems with data security. One can see, however, that some modifications are necessary because a straightforward encryption would guarantee data security but not data integrity. Note that for asymmetric encryption the encryption key is public, while the decryption key is private. Thus while nobody could read the encrypted data (nobody can decrypt), anybody can replace the original data by "new" data using the public encryption key. In the case of symmetric encryption the malicious user could not place correctly encoded data in the file but anything would do for subverting the data integrity. There are a number of ways to ensure that any attempt will be detected, for instance, inclusion of check bits in the plaintext which is to be encoded. Owing to the assumption about the probability of any symbol to appear in the ciphertext being flat, it is most unlikely that this can be detected and identified even if these check bits are in specific places in the plaintext. Another way is to employ two keys to encrypt the data, one encrypt key (K_A; public) and one decrypt key (L_B; private), i.e., instead of the data M one stores

$$E(D(M, L_B), K_A)$$

To retrieve M one first applies D with key L_A (private) and then E with key K_B (public). If both $E(D(M, L_B), K_A)$ and $D(M, L_B)$ are stored, any attempt at the data integrity of the file will be detected. Note that in any case it is not possible to *prevent* attempts at the data integrity. This appears to be an inherent limitation of any encryption scheme used to guarantee data security.

The main problem connected with encrypted data is that we cannot perform the operations which one can apply to the plaintext data, from arithmetic operations (addition, multiplication, e.g.) to sorting and searching (comparisons), to pattern matching (finding substrings, etc.). It turns out that in some cases this problem can be solved by using so-called privacy homomorphisms, due to Rivest, Adleman, and Dertouzos. Privacy homomorphisms are functions which allow to perform some of the desired operations directly on the encrypted data. We will first indicate the setting which is assumed and define these homomorphisms.

Then we show some inherent limitations, i.e., we demonstrate that certain operations cannot be performed if we want to maintain the cryptosecurity of the data. Finally, we give an example of a privacy homomorphism.

Consider a small bank B which rents computing time from a time sharing company T. It is desirable to keep B's files at the computing facility T for otherwise the cost of communication between B and T would be very high. However, if B's files are stored at T the problem arises that T's operating system may be subverted by an outsider in order to gain access to B's files. Even if this is not possible the systems programmers at the computing facility will always have access to B's files. Thus the confidential bank records would not be sufficiently secure. In order to attain an acceptable level of security the bank decides to encrypt all data before sending them off to T; this can be done using intelligent terminals, i.e., in a local (secure) environment, outside of T's influence. In other words, B will only keep encrypted information at T; therefore the security of its records is maintained. However, the cost of this setup is high, because it diminishes the usefulness of the computer drastically. For example, it is now impossible to get directly an answer to questions like:

- What is the total dollar amount of all outstanding loans?
- What is the dollar amount of all loans due within the next 12 months?

While before these computations could be done directly at the time sharing company T, now all the files involved must be sent back to B where the data must be decrypted, the desired information is extracted, and the necessary computations are performed. Thus the encryption of the data is not very useful to the bank. Or is it?

Let us consider what the real problem is. Assume we want to add two numbers, a and b; this will yield a third number c, $c = a + b$. Now suppose that a and b are encrypted, i.e., we are dealing with the numbers

$$E(a, K) \qquad \text{and} \qquad E(b, K)$$

for some key K. It would be nice and simple if we could just add up $E(a, K)$ and $E(b, K)$ and get as result $E(c, K)$. For if this is the case we do not have to send the data back, decrypt them, and only then perform the operation $+$. Rather we can directly apply the operation $+$ to the encrypted data and get the correct result back. In other words, we would like for E to behave as follows:

$$E(a, K) + E(b, K) = E(a + b, K)$$

But this is a well-known mathematical property; it simply signifies that E is a homomorphism with respect to the operation $+$. This shows that under certain assumptions concerning the encryption function we may be able to operate directly on the encrypted data after all. In the following we will make this idea more precise.

In our example with the bank the obvious operations involved are arithmetic operations, like $+$ and $*$. Also used may be predicates such as comparisons, like equal or less than. Operations and predicates are applied to elements of some set. Finally, we will have a number of distinguished constants in this set, like 0, 1, or 3.13149. Formally such a system is called an algebra,

$$(S;\ f_1, f_2, \ldots;\ p_1, p_2, \ldots;\ c_1, c_2, \ldots)$$

where S is the set of elements, the f_i are the operations, the p_i are the predicates, and the c_i (all in S) are the constants of the algebra. More specifically, this is the algebra of the plaintext data. Now we would like to have another algebra, the algebra of encrypted data such that to each element s in S there exists an element s' in the set S' of encrypted ele ments, to each operation f_i corresponds an operation f_i', and to each predicate p_i there is a predicate p_i', i.e.,

$$(S';\ f_1', f_2', \ldots;\ p_1', p_2', \ldots;\ c_1', c_2', \ldots)$$

is the algebra of encrypted data such that the algebra of plaintext data is homomorphically mapped by E onto the algebra of encrypted data, and this latter algebra in turn is mapped homomorphically by D onto the former algebra. That is:

(a) $E(S, K) = S'$ and $D(S', L) = S$, in particular, $E(c_i, K) = c_i'$ and $D(c_i', L) = c_i$ for all constants.

(b) If the operation f_i has arity m and $a_1, \ldots, a_m$ are elements of S, then

$$E(f_i(a_1, \ldots, a_m), K) = f_i'(E(a_1, K), \ldots, E(a_m, K))$$

Similarly, if f_i' has arity m and $a_1', \ldots, a_m'$ are from S', then

$$D(f_i'(a_1', \ldots, a_m'), L) = f_i(D(a_1', L), \ldots, D(a_m', L))$$

(c) If the predicate p_i has arity m and $a_1, \ldots, a_m$ are from S, then

$$E(p_i(a_1, \ldots, a_m), K) = p_i'(E(a_1, K), \ldots, E(a_m, K))$$

Similarly, if p_i' has arity m and $a_1', \ldots, a_m'$ are from S', then

$$D(p_i'(a_1', \ldots, a_m'), L) = p_i(D(a_1', L), \ldots, D(a_m', L))$$

Of course these conditions must hold in addition to the requirements on the encrypt function E and the decrypt function D set out in previous sections. However, we must impose a further condition, namely, that it is impossible to derive an efficient computation of D from the correspondence between the operations f_i and f_i', the predicates p_i and p_i', and the constants c_i and c_i'. That this condition can easily be violated is shown below. Indeed, it is the problems derived from violations of this very condition which seriously limit the applications of the otherwise very useful notion of privacy homomorphisms.

We will now give two examples which indicate some of the inherent limitations of privacy homomorphisms. They are both related to the last condition mentioned above and furthermore involve the use of comparisons. In the first example our algebra of plaintext data is over the natural numbers $N = \{0, 1, 2, \ldots\}$, we have one operation, namely, plus ($+$), one predicate, namely, less than ($<$), and one distinguished constant, namely, one (1):

$$A_p = (N;\ +;\ <;\ 1)$$

Assume now that E and D are privacy homomorphisms, i.e., A_p is mapped by E onto the algebra of encrypted data A_e,

$$A_e = (W;\ +';\ <';\ 1')$$

where W is some set, $+'$ is a binary operation on W, $<'$ is a predicate, and $1'$ is an element of W. We claim that D can easily be broken, that is, the privacy homomorphism is not secure. This can be seen as follows: Suppose we want to determine $D(x, L)$ for x some arbitrary element of W. We know $D(1', L) = 1$ since D is a homomorphism. If

$$x <' 1'$$

is true then we know that

$$D(x, L) = 0$$

Otherwise we compute

$$E(2, K) = 1' +' 1'$$

If

$$x <' E(2, K)$$

then we can deduce that

$$D(x, L) = 1$$

Otherwise we determine $E(4, K) = E(2, K) +' E(2, K)$, $E(8, K)$, $E(16, K)$, ... until we find $E(2^s, K)$ such that

$$x <' E(2^s, K) \text{ but not } x <' E(2^{s-1}, K)$$

This shows that

$$2^{s-1} \leq D(x, L) < 2^s$$

Now we add (by $+'$) the encrypted values of smaller powers of 2 (which we have already computed!) to $E(2^{s-1}, K)$ until we find a y such that $x <' E(y, K)$, etc. This is very similar to a binary search strategy. It follows that the precise decoding of x can be found in at most

$$2 \log(D(x, L))$$

additions ($+'$) and equally many comparisons ($<'$). Thus there is no secure decryption function for algebras of this type. It should be clear that this example easily generalizes to algebras where we can determine the encoded version of arbitrary constants and where there is a total order similar to $<$ or $\leq$.

The second example has again the natural numbers N as carrier of the algebra; furthermore, there are the operations addition $+$ and multiplication $*$ as well as the binary equality predicate $=$. Again we cannot have a secure privacy homomorphism. The argument is as follows. It is clear that for any nonzero number x we have

$$x = k \text{ iff } x^*x = \underbrace{x + \cdots + x}_{k \text{ times}}$$

Clearly this can be easily tested in the encrypted algebra if we have the corresponding operations $+'$ and $*'$ and the predicate $='$. Now, to detect the case where $x = 0$ all we must test is whether $x' =' x' +' x'$, as for any natural number x,

$$x = 0 \text{ iff } x = x + x$$

We conclude with an example, namely, we show that the encryption function proposed by Rivest, Shamir, and Adleman described in Section 4.3.2 is in fact a privacy homomorphism. More specifically, let Z_p be the integers modulo the prime p. Let $*_p$ denote multiplication modulo p, and let $=_p$ denote equality modulo p. Define n to be the product of p and q, $n = pq$, for q a prime about as large as p. Supposing that n is difficult to factor we define

$$E(x, e) = x^e \bmod n$$

Since

$$(x^e)(y^e) = (xy)^e$$

this is a homomorphism. Thus the encryption function from the algebra

$$(Z_p; *_p; =_p)$$

to itself is in fact a secure privacy homomorphism, secure by the arguments given in the last section and the observation that the condition of this section is satisfied. For further examples we refer to the original paper.

Bibliographic Note

Simmons[89] is again a good survey. The notion of a public cryptosystem is from Diffie and Hellman,[26,27] its first complete implementation can be found in Rivest *et al.*[80] A variant for which code breaking is *provably* equivalent to factoring is described in Rabin.[73] The second, well-known implementation which is based on the knapsack problem appears in Merkle and Hellman.[63] The NP-completeness of the knapsack problem and of other problems can be found in Karp.[49] In Shamir and Zippel[88] a method is given which shows that additional knowledge of m allows the determination of A with high probability when given B. References to approximation algorithms for the knapsack problem (which imply conditions for the choice of B) can be found in Merkle and Hellman.[63] The notion of a privacy homomorphism appears in Rivest *et al.*[79].

Exercises

1. Consider the Rivest–Shamir–Adleman scheme. Let $p = 31$ and let $q = 29$. Let $e = 11$.

   (a) Determine an integer d such that $ed = 1 \bmod ((p-1)(q-1))$.

   (b) Encode the message 876.

   (c) Assuming 876 is the transmitted cipher text, decode this message.

2. Explain in detail what happens if M is a multiple of either p or q. How likely is this to happen? Relate this to the difficulty of factoring!

3. Let $A = (1, 2, 4, 8, 16, \ldots, 2^{n-1})$.

   (a) Prove that (A, a) can be solved iff $0 \le a \le 2^n - 1$.

   (b) Show that the solution vector for the problem (A, a) is precisely the representation of the integer a in binary.

4. Assume that the knapsack vector A satisfies

$$a_i > a_1 + \cdots + a_{i-1} \qquad \text{for } i = 2, \ldots, n$$

   (a) Give a linear-time algorithm for determining whether (A, a) has a solution; furthermore, if there is a solution, the algorithm should produce it (in linear time).

   (b) Show that (A, a) has at most one solution for any integer a, i.e., if there is a solution it is unique.

5. Consider the knapsack vector

$$A = (14,\ 28,\ 56,\ 82,\ 90,\ 132,\ 197,\ 284,\ 341,\ 455)$$

   Show that $(A, 515)$ has three solutions, while $(A, 516)$ has none.

6. Let $A = (5, 10, 20)$, transform it to $B = (38, 29, 11)$ using $(w, m) = (17, 47)$, and then transform B to $C = (25, 87, 33)$ using $(w', m') = (3,89)$. Prove that there do not exist integers w'' and m'' such that $\mathrm{GCD}(w'', m'') = 1$ and (w'', m'') transforms A in one step to C.

7. Consider the algebra $(N; +, *; =)$ as discussed in Section 4.3.5. What is the complexity of compromising any privacy homomorphism from $(N; +, *; =)$ to any other algebra $(N'; +', *'; =')$?

8. Consider $p = 31$, $q = 29$, and $e = 11$ as in Exercise 1.
   (a) Encode the message 25.
   (b) Encode the message 19.
   (c) Encode the message 475 and compare the result with the product (modulo 899) of the encodings for 25 and 19. Do you have any explanation?

4.4. Authentication and Digital Signatures

4.4.1. Introduction

This fourth section on cryptosystems is concerned with the issues of authentication. As we pointed out in the Introduction to this chapter this aspect of cryptosystems is becoming more and more important as the number of distributed computer networks, electronic mail, and similar communication and/or distributed systems increases. There is no doubt about the significance of the problem involved. Our contention is that authentication and its more sophisticated version of digital signatures are problems which can be elegantly solved by cryptography. The most immediate authentication problem is the log-on procedure on a computer facility. The conventional method employs a password approach; the user sends the password and the system verifies it by comparing it against all other valid passwords. Given the fact that operating systems are notoriously unsafe it is very possible that a malicious user may obtain access to the file where all the passwords are stored. This very real danger can be avoided by storing only the encryption of the passwords (see previous section). However, this still leaves the possibility of a malicious user eavesdropping on the transmission line over which the password is communicated to the computer. Thus it should be clear that a solution is not entirely obvious.

An even more complicated situation exists with respect to digital signatures. Here we have all the previous problems but additionally the signature must also depend upon the message which is to be signed. If this is not the case, one genuine signature could be reused for many forged messages.

The general principle of authentication is to encrypt a rapidly changing unique value using a previously agreed upon key. An outline of a possible authentication sequence between users A and B follows; we assume the asymmetric encryption method, i.e., B knows A's encryption key K_A but only A knows the decryption key L_A.

1. B sends to A in cleartext a unique random data item, for example, the current time of day as known to B. (This is not entirely random but suffices for our purposes; note that even a small discrepancy may invalidate the attempted authentication.)

2. A encrypts the received time of day using L_A (secret) and sends this to B.

3. B decrypts A's message using K_A (public) and compares it with the time of day at which B received the message. If there is no long delay B is satisfied that A is the originator of the message.

The reason that B is satisfied is that only A can know L_A, and as we assume that it is computationally impossible to generate a key L_A' such that

$$E(D(M, L_A'), K_A) = M$$

this is a reasonable assumption.

Before presenting any specific implementations we will summarize the properties we expect from any acceptable digital signature SIG. This will be done in terms of communication from a sender S to a receiver R; hence the properties will not be symmetric in S and R.

(1) Given any message M, a signature SIG(S;M) must be a locally injective function from messages to signatures. A function f is called locally injective if $f(w) \neq f(w')$ for w and w' different but very similar. Clearly, the set of messages is usually larger than the set of signatures; this excludes an injective function which would be ideal. Injective functions can only be used if the signatures are at least as long as the messages; this is not very practical. However, minimally one must require that changing, for instance,

"$100,000.-" to "$900,000.-" or "1981" to "1991"

will also result in a change of the signature.

(2) Only S can produce SIG(S; M) for any given message M. If this property is violated, S's signature can be forged.

(3) If R receives a pair (M, SG) and SG is claimed to be S's signature of M, then R can verify without undue effort whether SG = SIG(S; M).

(4) For any given pair (M, SG) where R claims that SG = SIG(S; M), S can demonstrate whether this claim is valid or not.

In addition to these requirements we will postulate that the operations involved in producing and validating signatures be efficiently executable.

4.4.2. Signatures and Public Key Cryptography

In this section we outline how all these requirements can be satisfied if we use an elegant scheme which assumes an implementation (in fact, within limits *any* implementation) of the abstract notion of public key cryptography as proposed by Diffie and Hellman.

Suppose the participant A of a public key cryptosystem wants to send another participant B a signed message M. This can be achieved as follows.

A sends B the message

$$C = E(D(M, L_A), K_B)$$

where as usual K_B is B's public encryption key and L_A is A's secret decryption key. B retrieves M from C by first applying D with L_B (private) and then E with K_A (public).

By extension of the above argument B can be satisfied that this message came from A as only A knows L_A and thus can produce

$$D(M, L_A)$$

Note that there is no "physical" signature but the evidence that A actually signed M is compelling. In fact, by saving the message C, the user B can actually supply a *proof* that A signed M!

It is easily verified that requirements (1), (2), and (3) above are satisfied. However, in the scheme as it now stands requirement (4), which protects against repudiation, is not met. To restate this problem, assume that A sends a signed message C to B, for instance:

$$C = E(D(M, L_A), K_B)$$

It is conceivable that at some later point in time, A claims not to have sent this message to B. Worse yet, A may deliberately lose the secret key L_A thereby invalidating all signatures produced by A. One way to prevent this in the public key cryptosystems is by way of protocols. At the time when user A submits the public key K_A to the key distribution manager, A explicitly agrees to be responsible for all signatures produced with L_A, A's secret key. Note that K_A determines L_A (in fact, in the Rivest–Shamir–Adleman scheme uniquely). Should A really lose L_A, A agrees to notify immediately the key distribution manager. Furthermore, every signed message must be dated and time stamped; the recipient will not accept a signed message if there is a discrepancy between the time stamp and the actual time when the message was received.

These general remarks can be directly applied to the Rivest–Shamir–Adleman scheme. Rabin's modification also permits the generation of digital signatures. The same holds for the Merkle–Hellman method based on the knapsack problem; for the sake of completeness we briefly outline how to do this. If a sender S wants to transmit a signed message M to a recipient R, S uses his/her public trapdoor knapsack vector C and computes a vector X such that

$$c_1x_1 + \cdots + c_nx_n = M$$

This can be done (in principle at least) as S knows the simple knapsack vector A and the pair (w, m) which together produce C. Then S sends this vector X as the signed message to R. Clearly R can determine the message M as C is public. Also note that R (or anybody else other than S) is *not* able to compute X from C [S would not be able, either, except that S knows of course A and (w, m)!]. Thus the signature must be authentic. There is, however, one problem with this approach and this is the fact that M is given, i.e., there is no guarantee that the equation

$$C^*X = M$$

has a solution in X. Note that if there is no solution for a particular M, this message could not be signed. Thus, whether this signature method works depends on the density of the solutions, i.e., the fraction of numbers M between 0 and $c_1 + \cdots + c_n$ which have a solution to $C^*X = M$. If one starts with the vector $(1, 2, 4, \ldots, 2^{n-1})$ for A, the density will be high, however, for reasons of cryptosecurity this vector may not be acceptable. We refer to the original paper by Merkle and Hellman for further discussion of this issue.

All the preceding remarks are formulated in terms of asymmetric encryption. However, authentication and digital signatures can also be performed within the framework of symmetric encryption. In the following section we describe such a realization. Its main advantage over the more elegant and succinct public key signatures is the ability to discredit very transparently and within the method employed any attempt at cheating. Note that the above proposal of a defense against repudiation rests upon an external agreement which is not part proper of the signature system.

4.4.3. Signatures Based on Probabilistic Logic

In this section we outline a method for generating signatures which is closely related to symmetric encryption methods. We start with an encryption function E which maps pairs of binary k-tuples to binary k-tuples, for k some fixed integer,

$$E : K \times K \to K \qquad \text{where } K = \{0, 1\}^k$$

If x is a key in K and w a word in K, then $E(w, x)$ is the encoding of w by x. Since a message M may be longer than k, we split M into blocks of k bits each:

$$M = w_1 w_2 \ldots w_n, \qquad n \geq 1$$

E is extended to messages as follows:

$$E(M, x) = E\big(E(w_1 w_2 \ldots w_{n-1}, w_n), x\big) \qquad \text{for } n \geq 2$$

The function E may be any encryption function, symmetric or asymmetric, which satisfies three conditions:

(1) Given x and w in K, $E(w, x)$ can be rapidly computed.

(2) Given some sequence of pairs

$$\big(w_i, E(w_i, x)\big) \qquad \text{for } i = 1, \ldots, s$$

it is intractably difficult to produce another pair

$$\big(w', E(w', x)\big)$$

such that $w' \neq w_i$ for all $i = 1, \ldots, s$.

(3) Given any key x, it is intractably difficult to produce different messages M and N such that

$$E(M, x) = E(N, x)$$

Now we can summarize how signatures can be generated and verified in this scheme.

Suppose A and B wish to conduct digitalized signed correspondence over an insecure channel. Each of them chooses H keys, namely, $x_1, \ldots, x_H$ for A and $y_1, \ldots, y_H$ for B. Both sequences must be kept secret. A then computes a sequence,

$$X = (X_1, \ldots, X_H) \qquad \text{where } X_i = E(x_i, f(i)) \text{ for } i = 1, \ldots, H$$

and B computes a similar sequence,

$$Y = (Y_1, \ldots, Y_H) \qquad \text{where } Y_i = E(y_i, f(i)) \text{ for } i = 1, \ldots, H$$

where $f(i)$ is the number i in binary notation (k bits). Thus $f(i)$ is in K. A then passes the sequence X on to B and B passes the sequence Y on to A.

If A wishes to send B a signed message, A transmits the pair

$$(M, \text{SIG}(\text{A}; M))$$

to B where SIG(A; M) is A's signature of message M. M can be any string of words; in particular, M can be encrypted. The signature SIG(A; M) is a sequence of T strings in K which are defined as follows (T must be even):

$$\text{SIG}(\text{A}; M) = E(x_{i+1}, E(M, f(0))), \ldots, E(x_{i+T}, E(M, f(0)))$$

Upon receipt of $(M, (v_1, \ldots, v_T))$, B verifies that $v_1, \ldots, v_T$ is in fact A's signature of M in the following way: B randomly chooses $T/2$ integers $i_1, \ldots, i_{T/2}$ between $i + 1$ and $i + T$ and requires A to divulge the keys x_t for $t = i_1, \ldots, i_{T/2}$. B can now verify whether these are the keys initially agreed upon by computing $E(x_t, f(t))$ and comparing it to X_t, for $t = i_1, \ldots, i_{T/2}$. B also checks that $v_{t-1} = E(x_t, E(M, f(0)))$ for all these t. Only if all tests are positive, B accepts the message as signed A. This follows from the observation that in the given situation it is extremely improbable that anybody else could have produced this signature. Thus B can verify whether the signature is genuine. We remark that each key is to be used only once; thus X and Y are split up into contiguous blocks of T keys each, and every block is to be used to sign exactly one message. Consequently H should be a multiple of T.

It follows that for an acceptable encryption function E, the requirements (1) and (2) set out above are clearly satisfied. The technique just outlined also ensures that (3) is met. Finally, we must make certain that repudiation is not possible. If we manage to do this then all four requirements are met and hence we have a useful digital signature.

We claim that at the cost of a somewhat more complicated exchange protocol it is possible to protect against repudiation. This can in fact be done without any permanent central authority. Suppose A sends a pair

$$(M, (v_1, \ldots, v_T))$$

to B. B verifies that $(v_1, \ldots, v_T) = \mathrm{SIG}(\mathrm{A}; M)$. If this test is positive, B sends the pair

$$(N, (w_1, \ldots, w_T))$$

to A and A verifies that $(w_1, \ldots, w_T) = \mathrm{SIG}(\mathrm{B}; N)$. If this is not the case A informs B that M is rejected, otherwise A verifies that

$$N = \$\, M\, \mathrm{SIG}(\mathrm{A}; M)$$

where $\$$ is a special token in K which indicates that what follows is the previous message sent from A to B. Again if this is not the case A rejects M. If $N = \$\, \mathrm{M}\, \mathrm{SIG}(\mathrm{A}; M)$, then A sends

$$(P, (z_1, \ldots, z_T))$$

to B and B verifies that $(z_1, \ldots, z_T) = \mathrm{SIG}(\mathrm{A}; P)$ and

$$P = \sharp\, N\, \mathrm{SIG}(\mathrm{B}; N)$$

where $\sharp$ is another special token in K indicating that what follows is confirmation of the original message M sent from A to B, i.e., the *last* message A sent to B. If either of these tests fails B finally rejects M; otherwise B accepts M. Note that in this scheme each participant will mark $2T$ keys as used in order to ensure that no old keys can be reused.

With this protocol it can be verified that all four requirements are satisfied by our method. For more details we refer to the original papers.

Bibliographic Note

The first proposal of digital signatures based on asymmetric encryption (public key cryptosystems) is contained in Diffie and Hellman.[26,27] The two most important implementations of public key cryptography can both be used to produce digital signatures (Rivest *et al.*[80] and Merkle and Hellman[63]). Rabin's method[73] also allows the generation of digital signatures. Symmetric encryption methods can also be used to authenticate messages (see, for instance, Feistel[31]). The method described in the last section is taken from Rabin[74] and Leiss.[54]

4.5. Protecting Ownership of Proprietary Data and Software

4.5.1. Introduction and Motivation

In this chapter we will discuss some applications of cryptography, namely, protection of software against theft and other issues related to proprietary rights on products which can be perfectly copied.

Some of the most dynamic and innovative talent is without doubt concentrated in industries related to computers, in particular, software houses. While the need for their products, program packages, systems, and their derivatives, is certainly a crucial factor for this, so is the medium itself, which allows the transition from the idea to the final product with speed and ease unprecedented and unparalleled in any other area of production. Consequently this market has become economically very significant, in excess of one billion dollars per year. Faced with such a volume of business the question of theft of the products comes up quite naturally—or at least it would were it a different, more tangible, product. As it is, the legal profession does not have a clear understanding of the product involved, much less of the question of what constitutes theft thereof. Even worse, until recently the data processing community itself did not appear too concerned about the problem. Only in the last few years did some of this concern become apparent in the open literature as well as in the popular press.

In this section we address two problems related to theft of software and its derivatives. The first aims at preventing theft; consequently it must achieve this objective entirely by technical means. Legal procedures at best act as deterrent; they can never *prevent* criminal actions. The second has to do with detection. Note that in the case of a copied program it is impossible to detect in the usual way that it was copied. This is due to the fact that every copy is a "perfect copy". Compare this situation with a "copier", i.e., forger of bank notes: It would be virtually impossible to become alerted to the fact that bank notes are forged (unless the forger overdid it and printed them in noticeably large quantities) if the forgeries were perfect copies of the originals. It should be clear that without detection there can hardly be conviction. Thus if one cannot prevent theft of software in the first place, it is desirable to be able to detect it as a necessary condition (alas not sufficient!) for prosecution.

Now a statement of the two scenarios. Both of them have in common that a producer P, which may be a software house, a data collection agency, or a forecasting bureau, produces a certain product X. Of this product we will only assume for the moment that it can be digitalized and electronically transmitted. This ensures that perfect copies of X can easily be made. Examples are any kind of software, but also numerical data or any kind of information. The reader should note that there is a fundamental difference between a program and a table of statistics. A program is virtually worthless unless it can eventually be executed (cf. a program using the full power of ALGOL68). (We take an engineer's approach to the aesthetic qualities of a program; in particular we will always prefer an "ugly" program which works to a "beautiful", "aesthetically pleasing" one which "almost" works.) Thus a program's primary objective is to "do something". This action by itself might be employed for preventing or detecting theft of X. (We will assume that the programs in question are so complicated that it is not immediately apparent what the output is for any given input.) On the other hand, a statistical table may be valuable in itself; no other medium such as a compiler is required. This difference is very important for our purposes; for this reason we will use the term *dynamic information* for software and *static information* for statistical tables.

At this pont the alert reader may have already noticed that the notion of preventing theft of information is somewhat illusory; at least in the case of static information it is impossible to prevent theft once the information is revealed. Clearly there is no way (legal anyway) to (unconditionally) *prevent* the recipient of the information from making a copy of it (or to remember it!) and pass it on to others. However, on second thought, no safety box is absolutely safe either, nor is any bank vault or Fort Knox for that matter. What is true is that an attempt at breaking the security in these cases is difficult. "Difficult" of course is a relative term; the degree of difficulty must be in relation to the value of the property to be protected. Ideally we want to create a situation where the cost of breaking the security of a system exceeds the advantage which can thereby be derived. Formulating this in terms of information, we would like to have a situation where successful copying requires as much knowledge, expertise, and work from the copier as deriving the information directly (writing the program from scratch or computing the statistics from data in the public domain).

The remainder of this chapter is organized as follows. In the next section we will discuss protection of dynamic information. We will concentrate on a method for prevention of theft which employs encryption. Its disadvantage is that it requires hardware modifications. The last section is concerned with static information. Both prevention and detection are issues; however, for static information there are no methods which would achieve the objectives with a high degree of confidence.

4.5.2. Dynamic Information

Suppose a buyer B buys a program X from a software house S. In almost all cases (discount turn key systems) B will insist that S furnish the source code of X. In fact, B will want to get (at least) a well-documented source code plus an explicit description of the algorithms or methods used and of the input–output behavior of X. B has good reasons for this demand. First of all, it is very unlikely that a program developed at one installation will immediately run on another installation even if both have the same central computer. Virtually always some modification is required. Secondly, no large piece of software is entirely correct; many errors and inconsistencies do not significantly affect its performance in general, although they may show up later in certain cases. These errors are customarily removed sometime after they were detected. This implies that ongoing maintenance is required. In some cases the software house is willing and able to perform this; in many cases it is not. Furthermore what is the buyer to do if S ceases to exist or discontinues its maintenance of X? Thirdly, it can be safely assumed that B will want to customize the program, i.e., tailor it to its specific needs and/or remove unnecessary parts from it. This is virtually impossible without the source code. On the other hand, software houses usually do not do this for their clients. For these reasons it is customary in the software industry to reveal the source code to the buyer. However, this leaves the vendor wide open to the threat of theft as the buyer has now complete control over the product X and in turn could sell copies of X to other interested parties, presumably at a lower price than S.

In Demillo *et al.*[20] a partial solution to the problem of *detecting* theft is described. It consists of "blurring" the comparisons in the program to be protected by superimposed coding and interspersing the original code with "false", i.e., meaningless, statements. On the one hand, this allows the user to some extent to modify the program; on the other hand, it requires a great deal of work to determine the precise meaning of the comparisons (more than cubic time in the number of independent program parameters). Unfortunately, this scheme *does not* prevent the buyer B from reselling the program as it is! In other words, B will not be able to determine which comparisons the program makes and consequently B will not be able to rewrite the program in such a way that a careful comparison with the original would not reveal conclusive similarities, but B *will* be able to simply sell it as is! Since software houses usually are not very large but the number of interested parties may be, it is quite unreasonable to expect that such theft is detected. What is really required is a method of *preventing* theft from the very beginning.

In the following we restrict our attention to software which processes a substantial amount of (not necessarily numerical) data and contains a large number of comparisons and operations involving the data and constants. Typically in this kind of software, there is only a limited number of functions which moreover must be performed frequently and by a large number of enterprises in a very similar way. An example is a payroll program which must be run at least once a week and is basically the same for many firms; thus it is a good candidate for theft. Another example might be a compiler for a commonly used programming language like COBOL or FORTRAN. The solution we propose is conceptually quite simple. Its implementation will require substantial hardware modifications. While this is probably viable only for relatively large computing centers, these are exactly the installations from which a software house would derive substantial income; usually small centers will not pay large sums for such programs.

The usual way of ensuring that some information cannot be divulged is by encrypting it using some key K and keeping K secret. This approach is clearly not useful as the buyer has legitimate reasons to obtain the source code (naturally, in unencrypted form). Nevertheless we intend to use encryption, albeit in a different way.

In the following we will denote the original program by X; the program modified as described below will be denoted by X′. The buyer B will receive X′; the objective of the modification is to make it difficult for B to restore X, given X′. The main idea is that all constants appearing in X′ will be encrypted; that is, if X contains a constant, X′ will contain this constant only in encrypted form. The central processing unit would have to be modified to include a *secure crypto unit*. The technical setup would be comparable to the addition of a floating-point arithmetic unit. This unit would perform the usual arithmetic operations and comparisons but with the difference that the constants of the program are encrypted. Thus, it first decrypts the program constant before it performs the operation. Furthermore the decryption method is dependent upon the operator or comparison in which the constant is involved (we will say the encryption is context dependent). More specifically, the crypto unit would use *one* decryption algorithm but *several* different keys, one for each operation and comparison. For example, if

$$\text{CONST} + \text{SUM}$$

is to be performed the key K8 is used to decrypt the constant CONST, and if

$$\text{CONST} \geq \text{SUM}$$

is to be performed the key K0 is employed before the operation (or comparison) is executed. In fact, if higher security is required the key can depend on the name if the variable which appears as the other operand in the statement.

A crypto unit is secure if only the input–output behavior can be observed and if it is impossible to determine the various keys (even if physical force is applied).

The program X′ is then obtained as follows. The original program X is scanned for constants. Every constant encountered will be encrypted; the choice of the encryption keys depends on the context of the constant. Thus the resulting program, say, Y, when executed with the secure crypto unit has the same input–output behavior as X. Additionally, Y will be padded with meaningless statements such as

$$\textbf{if } A \geq E(0) \textbf{ and } A < E(-1) \textbf{ then } A := A + E(3)$$

whereby $E(i)$ denotes the encryption of i corresponding to the particular context of the constant i. Consequently it will be far from obvious that these statements are in fact meaningless and do not affect the behavior of the program. A similar approach is taken in Demillo *et al.*[20].

In order to be useful this method must have two properties: It must allow the buyer to modify (within limits) the program X′ to suit it to the specific requirements the buyer might have, and it must ensure that theft of the program, reselling it to another interested party, is greatly complicated. Obviously we will have to assume that detailed in-line comments are removed; general comments such as those stating briefly the input–output behavior of subroutines, or explaining the function of parameters, global variables, and local variables, may be retained.

We first comment on the ability to modify such programs. This is not entirely trivial as it will usually not be possible to identify the precise point where a certain modification should best be made. (Incidentally, this very same problem occurs in the scheme in Demillo *et al.*,[20] too.) For example, if a firm wants to give a special bonus to all employees born before its founder's date of birth it is unlikely that a general payroll program will accommodate this. The applications programmer in this case would have to insert a procedure whose only function is to test for this condition and to make the proper adjustments. A competent applications programmer should be able to find a place in X′ where this procedure can be safely called; note that care must be taken in placing it as the meaning of the statements in X′ is not completely known (they may even be meaningless).

We now come to the contention that this method will actually make theft substantially more difficult. First of all we observe that the program cannot be directly copied. This is due to the fact that the secure crypto unit cannot be duplicated as the keys are not known to the user. Consequently anybody who wants to steal the program for reselling it must first be able to determine the original constants in X, given the modified program X′. Note that the modified program without the secure crypto unit is meaningless as the comparisons and operations would be with completely unrelated constants. In order to be able to copy the program X, it is therefore necessary to decrypt the constants in the (modified) program X′. There are two possible ways of determining the constants in X; either one breaks the code, i.e., determines the keys (the encoding *method* can usually not be assumed secret), or one determines the constants directly. Breaking the code is a substantially more difficult problem than the direct determination of the (original) constants, which we discuss next: Once one has the key one can easily determine the constants but the converse is clearly not true. Determining the (original) constants of the program X in a direct way is a rather dangerous attack if one does not use some precaution. The problem here is that in a typical program many of the constants are between -10 and 10, and it can certainly be assumed that almost all constants lie between $-1{,}000{,}000$ and 1,000,000. Consequently, it is possible to feed these numbers one by one to the crypto unit until one of them results in the same behavior; this will eventually yield the original value of the encrypted constant. If comparisons are freely available within the crypto unit it may even be possible to employ a binary search strategy as described in Rivest *et al.*,[79] Parts II and IV. The last problem is avoided by encrypting with different keys in the context of different operations and comparisons. Another problem is of course that by inserting probes into the program (which is an explicitly permitted modification of the program X′!) the malicious user is able to determine the encryptions of some constants. For example, if there is in X′ the statement

$$A := A + E(5)$$

where $E(5)$ denotes the encryption of 5, it is possible to decrypt $E(5)$ simply by printing the value of A before and after execution of this statement and then subtracting the former from the latter value. How then are we going to attain any measure of security? The main observation to this end is that it is not obvious to determine the flow of control for a run of the program because one does not know the decryption of the constants involved in predicates (e.g., tests for equality, greater, etc.) and that there is no easy way to do this. At best if comparisons are freely available within the crypto unit one can perform a binary search for each of the constants. While this is clearly possible we maintain that for a large and complicated program this will involve a fair amount of work and will certainly deter some dishonest buyers, although admittedly not necessarily all. At any rate it is necessary to enlarge the range of the constants; this can be done by "splitting the constants"; for example, we have

$$1 = 10^{14}/(5^{14} \cdot 2^{14})$$

This will be in X′:

$$E(1) = E(10^{14})/(E(5^{14}) \cdot E(2^{14}))$$

To illustrate we contrast a program fragment in X with its counterpart in X′:

In X:

```
        .
        .
        .
        I := 1
L1:     if I = 100 then go to L2;
        A(I) := I;
        I := I + 1;
        go to L1;
L2:
        .
        .
        .
```

This program fragment might look as follows in X′:

```
        .
        .
        .
L0:     I := E(10^14)
        if I/E(5^14) = E(2^14) then I := E(1) else go to L2;
L1:     if I*E(10^10) = E(10^12) then go to L2;
        A(I) := I;
        I := I + E(1);
        if I*E(10^9) ≥ E(10^8) + I and I*E(0) ≥ I − E(101)
          then go to L1 else I := E(15);
        go to L0;
L2:
        .
        .
        .
```

The reader should keep in mind that the buyer of X′ will see only the encrypted constants. Already with this small fragment it is not entirely obvious to see what will happen; this holds even if we insert probes between all the statements. Another possibility is to involve constants in places of X′ where they are not required in X. For example, the simple assignment

$$A := B$$

could be changed to

$$A := (E(1)*B + E(0))*E(1)$$

which is of course far more complicated to understand. Note that even the first encoding of the constant 1 can be different from its second encoding if the whole context of the constant is taken into account (such as the name of the variables or expressions with which the constant is connected by the operation and on which side of the operator the constant appears).

It must be said that theoretically it will always be possible to determine all the constants and then restore the original program X. But we believe that this will require the thief to fully understand the program. Anybody with this knowledge and expertise could also write the complete program from scratch with no more effort than is required to copy it, and this would of course be perfectly legal.

In summary we proposed a method which will give some degree of protection against direct theft of dynamic information. We realize that the security of the proposed method is not very strong, but we believe that it will discourage theft to some extent. It should be kept in mind that up until now there is no method at all which will prevent theft; the method described in Demillo *et al.*[20] only assists in detecting theft but cannot be used for preventing it. Our method is conceptually simple but requires some initial modification of the hardware. The extent of this modification is similar to that of a floating-point arithmetic unit, i.e., it is a board which is connected to the CPU; in addition it must be physically secure and have some mechanism to receive keys. Changing the key and the encrypted constants will increase the security of the proposed method. This is not very difficult as the periodic maintenance can be done using conventional data links (e.g., telephone lines). Note that even the transmission of the new encryption key need not be secure as they may be sent encrypted and the secure crypto unit (in a separate component) could decrypt the transmitted keys to get the actual keys which are to be used in the decryption of the constants of the program. This separate component must employ a counter to ensure that retransmission of the same input (encrypted key) will result in *different* actual keys. The replacement of the encrypted constants can be done over insecure lines as they will be under the user's control anyway. Finally, the secure crypto unit could contain another counter ensuring that the key *must* be changed every once in a while; this would guarantee that the user could not intercept the changes transmitted by the software house in order to keep the same key longer, thus enabling the user to crack the code.

4.5.3. Static Information

We now turn our attention to static information such as statistical tables. The situation here is quite different from that of dynamic information in that by revealing it the supplier has no control over it at all. In fact, a dishonest user might just be able to remember the data and divulge it later on. It is clearly impossible to prevent this kind of theft. However, in many cases it would be already sufficient if one could protect oneself against *future* theft. Such a scheme will be outlined below. The protection is again relative because it can be circumvented but this requires a good deal of work. Furthermore, the amount of work increases with the number of copies the thief makes.

We start by describing the kind of situation we have in mind. There is a growing number of small to medium size firms which specialize in data collection, economics research, and forecasting; these firms act as consultants to other enterprises. In many cases the work of these firms involves handling large amounts of data; consequently their data processing utilizes computers. Most of the time the consultant rents computer services from a large center. As timeliness and speed of the responses are crucial in this business the data are frequently supplied online to the users of the service which the consultant provides. The user pays the computer center for the resources used in the usual way while the consultant is paid a flat fee, giving the user access to this service for a specified period. Note that frequently there is no differential in the price depending on the usage of the service, i.e., the consultant's fee is the same regardless of whether the buyer uses the service often or seldom. Also note that the data files may be periodically updated; thus copying the whole file once does not necessarily constitute copying of the consultant's services. In other words, continuous contact may well be required. The following problem arises: Suppose a user U acquires the service from the consultant and obtains the necessary authorization to use certain files and programs. This authorization will most probably be granted by way of passwords. U, having paid the fee and obtained the passwords, could now sell the passwords to another interested party V. U may be charged for V's computing and log-on time but this will be negligible compared with the price of the service. Thus the question is: How can the consultant prevent this abuse?

A relatively simple way of achieving this is by way of a variable password scheme. At the end of any session in which the service was used a password is issued to the user. Before being able to use the service in the next session the user must provide the password issued in the last session. It should be clear what happens if there is a user V other than the legitimate user U: The first time V uses the service with U's authorization (i.e., passwords) the system issues a new password. However, now U cannot use the service any longer as U does not know the new password. Assuming that the service is to be used frequently it follows immediately that U and V must constantly exchange the current variable password. While this may not prevent U from giving one user access to the service the necessary setup will soon become quite unfeasible if there are more additional users. Either all users must report back to one of them or else it would take many calls to find out who used the service last. Thus with a growing number of illegal users the logistics will become highly cumbersome. Note that this does not apply to legitimate users; if U and U′ are both legitimate users then both have independent variable password schemes.

A large number of independent password schemes can be obtained as follows. This method is quite simple; however, it should be kept in mind that the security of this method must be in relation to the security of the total organization. Since we already agreed that it is impossible to prevent theft of static information unconditionally, it is acceptable to employ a naive approach for the generation of variable passwords. Choose a random number generator R with period p, i.e., if $x(0)$ is the seed and $x(i+1) := R(x(i))$, then $x(j) \neq x(k)$ for all $0 \leq j < k < p$; p should be as large as possible (see, for instance, Knuth[50]). Suppose now that one needs M different password schemes (i.e., there are M legitimate users); then the first user will get as initial password the number $x(0)$, the second user will get as initial password $x([p/M])$, the third user $x(2[p/M])$, etc. Thus in any case the next password is simply $R(y)$ if y is the previous password. This should give very satisfactory results for M up to 10,000 as a good random number generator will have $p \gg 100{,}000{,}000$ (i.e., every user can have more than 10,000 sessions).

We close with an observation concerning detection. In the scenario in which we are interested in this case, it is not so important to know *that* some table has been copied but rather *from whom* it has been copied, the implication being that this user assisted in the copying. One way which has been used in the past to identify copies of mathematical tables such as sine or logarithm is to change some less significant digits. This scheme can be refined for our purposes. Each user gets a "personalized" copy of the data requested. This is achieved in the following way. Suppose the numbers are computed with s significant digits. Then the user will obtain numbers with $s + 1$ significant digits such that rounding to s digits will give the actually computed numbers. The sequence of last digits is chosen in such a way that it uniquely identifies the recipient, i.e., no two users will get the same sequence. As there are ten possibilities for any last digit (for example, .5 corresponds to any of .46, . . . , .49, .50, .51, . . . , .55), the probability that for m numbers this particular sequence was chosen by accident is $1/10^m$. The method has the drawback that it must not be known to the user; otherwise the user could first round, thereby removing the "incriminating personalization," and then pass the table on. In order to avoid this the producer of the table would have to apply this modification scheme to more significant digits; this may or may not be feasible. As we already pointed out above, protection of static information is very difficult.

Bibliographic Note

A discussion of issues related to detection of theft of software is contained in Demillo *et al.*;[20] this paper also contains a description of how superimposed coding can be employed to this end. The method for preventing theft is from Leiss.[58] The ideas contained in the last section on static information are mostly in the public domain.

References

1. Adleman, L. M. On distinguishing prime numbers from composite numbers, *IEEE Proc. FOCS*, 387–406 (1980).
2. Astrahan, M. M., *et al.* System R: Relational approach to database management, *ACM Trans. Database Syst.* **1** (2), 97–137 (June 1976).
3. Bell, D. E., and La Padula, L. J. Secure computer systems: Unified exposition and MULTICS implementation, The Mitre Corporation, ESD-TR-75-306, July 1975.
4. Berlekamp, E. R. Factoring polynomials over large finite fields, *Math. Comput.* **24**, 713–735 (1970).
5. Bishop, M., and Snyder, L. The Transfer of information and authority in a protection system, Department of Computer Science, Technical Report No. 166, Yale University, July 1979.
6. Brassard, G. Relativized cryptography, *IEEE Proc. FOCS*, 383–391 (1979).
7. Brassard, G. A note on the complexity of cryptography, *IEEE Trans. Inform. Theory* **IT-25**(2), 232–233 (March 1979).
8. Brassard, G. A time–luck tradeoff in cryptography, *IEEE Proc. FOCS*, 380–386 (1980).
9. Chamberlin, D. D., *et al.* Data base system authorization, in *Foundations of Secure Computation*, R. D. Demillo, D. Dobkin, A. K. Jones, and R. J. Lipton, eds., Academic Press, New York (1978), pp. 39–55.
10. Chin, F. Y. Security in statistical databases for queries with small counts, *ACM Trans. Database Syst.* **3**(1), 92–104 (March 1978).
11. Codd, E. F. A relational model of data for large shared data banks, *Commun. ACM*, **13**(6) 377–387 (June 1970).
12. Codd, E. F. A data base sublanguage founded on the relational calculus, in Proceedings of the 1971 ACM-SIGFIDET Workshop on Data Description, Access and Control, San Diego, California, November 1971, pp. 35–68.
13. Codd, E. F. Further normalization of the relational data base model, in *Courant Computer Science Symposium*, *Vol. 6: Data Base Systems*, R. Rustin, ed., Prentice-Hall, Englewood Cliffs, New Jersey (1971), pp. 33–64.

14. CODD, E. F. Relational completeness of data base sublanguages, in *Courant Computer Science Symposium, Vol. 6: Data Base Systems*, R. Rustin, ed., Prentice-Hall, Englewood Cliffs, New Jersey (1971), pp. 65–98.
15. CODD, E. F. Recent investigations in relational data base systems, in *Proceedings of the IFIP Congress 1974*, North-Holland, Amsterdam (1974), pp. 1017–1021.
16. DATE, C. J. *An Introduction to Data Base Systems*, Addison-Wesley, Reading, Massachusetts (1975).
17. DAVIDA, G. I., LINTON, D. J., SZELAG, C. R., and WELLS, D. L. Database security, *IEEE Trans. Software Eng.* **SE-4**(6), 531–533 (November 1978).
18. DEMILLO, R. D., DOBKIN, D., JONES, A. K., and LIPTON, R. J., eds. *Foundations of Secure Computation*, Academic Press, New York (1978).
19. DEMILLO, R. A., DOBKIN, D., and LIPTON, R. J. Even data bases that lie can be compromised, *IEEE Trans. Software Eng.* **SE-4**(1), 73–75 (January 1978).
20. DEMILLO, R., LIPTON, R., and MCNEIL, L. Proprietary software protection, in *Foundations of Secure Computation*, R. D. Demillo, D. Dobkin, A. K. Jones, and R. J. Lipton, eds., Academic Press, New York (1978), pp. 115–131.
21. DENNING, D. E. Are statistical data bases secure? Paper presented at the NCC in Annaheim, California, June 1978.
22. DENNING, D. E. A review of research on statistical data base security, in *Foundations of Secure Computation*, R. D. Demillo, D. Dobkin, A. K. Jones, and R. J. Lipton, eds., Academic Press, New York (1978), pp. 15–25.
23. DENNING, D. E., and DENNING, P. J. Data security, *ACM Computing Surveys* **11**(3), 227–250 (September 1979).
24. DENNING, D. E., DENNING, P. J., and SCHWARTZ, M. D. The tracker: A threat to statistical database security, *ACM Trans. Database Syst.* **4**(1), 76–96 (March 1979).
25. DENNING, D. E., SCHLÖRER, J. A fast procedure of finding a tracker in a statistical database, *ACM Trans. Database Syst.* **5**(1), 88–102 (March 1980).
26. DIFFIE, W., and HELLMAN, M. E. New directions in cryptography, *IEEE Trans. Inf. Theory* **IT-23**(6), 644–654 (November 1976).
27. DIFFIE, W., and HELLMAN, M. E. Exhaustive cryptanalysis of the NBS Data Encryption Standard, *Computer* **10**(6), 74–84 (June 1977).
28. DOBKIN, D., JONES, A. K., and LIPTON, R. J. Secure databases: Protection against user inference, *ACM Trans. Database Syst.* **4**(1), 97–106 (March 1979).
29. DOBKIN, D., LIPTON, R. J., and REISS, S. P. Aspects of the database security problem, in Proceedings of a Conference on Theoretical Computer Science, August 15–17, 1977, University of Waterloo, Waterloo, Ontario.
30. FAGIN, R. On an authorization mechanism, *ACM Trans. Database Syst.* **3**(3), 310–319 (September 1978).
31. FEISTEL, H. Cryptography and computer privacy, *Sci. Am.* **228**(5), 15–23 (May 1973).
32. FLYNN, M. J. *et al.*, *Operating Systems: An Advanced Course*, Springer Lecture Notes in Computer Science No. 60, Springer, New York (1978).
33. GAINES, R. S., and SHAPIRO, N. Z. Some security principles and their application to computer security, in *Foundations of Secure Computation*, R. D. Demillo, D. Dobkin, A. K. Jones, and R. J. Lipton, eds., Academic Press, New York (1978), pp. 223–236.
34. GRAHAM, G. S., and DENNING, P. J. Protection, principles and practice, in

Proceedings of the 1972 AFIPS Spring Joint Computer Conference, Vol. 40, AFIPS Press, Montvale, New Jersey (1972), pp. 417–429.

35. Griffiths, P. P., and Wade, B. W. An authorization mechanism for a relational database system, in *ACM Trans. Database Syst.* **1**(3), 242–255 (September 1976).
36. Hall, M., Jr. *Combinatorial Theory*, Blaisdell Publishing Company, Waltham, Massachusetts (1967).
37. Harary, F. *Graph Theory*, Addison-Wesley, Reading, Massachusetts (1969).
38. Harrison, M. A., and Ruzzo, W. L. Monotonic protection systems, in *Foundations of Secure Computation*, R. D. Demillo *et al.*, eds., Academic Press, New York (1978), pp. 337–365.
39. Harrison, M. A., Ruzzo, W. L., and Ullman, J. D. Protection in operating systems, *Commun. ACM* **19**(8), 461–471 (August 1976).
40. Hellman, M. E. A cryptanalytic time–memory trade-off, *IEEE Trans. Inf. Theory* **IT-26**(4), 401–406 (July 1980).
41. Herlestam, T. Critical remarks on some public-key cryptosystems, *BIT* **18**, 493–496 (1978).
42. Hopcroft, J. E., and Ullman, J. D. *Introduction to Automata Theory, Languages, and Computation*, Addison-Wesley, Reading, Massachusetts (1979).
43. Jones, A. K. Protection in programmed systems, Ph.D. thesis, Carnegie-Mellon University, 1973.
44. Jones, A. K. Protection mechanisms and the enforcement of security policies, in *Operating Systems: An Advanced Course*, Springer Lecture Notes in Computer Science No. 60, M. J. Flynn *et al.*, eds., Springer, New York (1978), pp. 229–251.
45. Jones, A. K. Protection mechanism models: Their usefulness, in *Foundations of Secure Computation*, R. D. Demillo *et al.*, eds., Academic Press, New York (1978), pp. 237–254.
46. Jones, A. K., Lipton, R. J., and Snyder, L. A linear time algorithm for deciding security, in *Proceedings of the 17th Annual Symposium on Foundations of Computer Science*, IEEE Computer Society (1976).
47. Kam, J. B., and Davida, G. I. A structured design of substitution–permutation encryption networks, in *Foundations of Secure Computation*, R. D. Demillo *et al.*, eds., Academic Press, New York (1978), pp. 95–113.
48. Kam, J. B., and Ullman, J. D. A model of statistical databases and their security, *ACM Trans. Database Syst.* **2**(1), 1–10 (March 1977).
49. Karp, R. M. Reducibility among combinatorial problems, in *Complexity of Computer Computation*, R. N. Miller and J. W. Thatcher, eds., Plenum Press, New York (1972), pp. 85–104.
50. Knuth, D. E. *The Art of Computer Programming*, Vol. II, Addison-Wesley, Reading, Massachusetts (1969).
51. Leiss, E. Security in databases where queries involve averages, Research Report CS-77-33, Department of Computer Science, University of Waterloo, Waterloo, Ontario, October 1977.
52. Leiss, E. Database security and restricted randomizing, Technical Report No. 63-79, Department of Computer Science, University of Kentucky, Lexington, Kentucky, July 1979; see also Proceedings of Primera Conferencia Nacional en Teoría de Computación y Desarrollo de Software, Santiago, Chile, August 22–24, 1979.

53. LEISS, E. The number of known elements required for global compromise with s-queries, in Proceedings of the Eleventh Southeastern Conference on Combinatorics, Graph Theory, and Computing, Boca Raton, Florida, March 1980.
54. LEISS, E. A note on a signature system based on probabilistic logic, *Inf. Process. Lett.* **11**(2), 110–113.
55. LEISS, E. Authorization systems with bounded propagation, in Proceedings of the 18th Annual Allerton Conference on Communication, Control, and Computing, Montecello Illinois (October 1980).
56. LEISS, E. Database security and the representation of graphs as query graphs, in Proceedings of the 12th Southeastern Conference on Combinatorics, Graph Theory, and Computing, Baton Rouge, Louisiana, March 1981.
57. LEISS, E. On the security of randomized databases: A simulation, Technical Report No. UH-CS-81-01, University of Houston, February 1981.
58. LEISS, E. Protecting ownership of proprietary data and software, Technical Report No. UH-CS-80-04, University of Houston, August 1980.
59. LIPTON, R. J., and BUDD, T. A. On classes of protection systems, in *Foundations of Secure Computation*, R. D. Demillo *et al.*, eds., Academic Press, New York (1978), pp. 281–298.
60. LIPTON, R. J., and SNYDER, L. A linear time algorithm for deciding subject security, *Journal ACM* **24**(3), 455–464 (1977).
61. LIPTON, R. J., and SNYDER, L. On synchronization and security, in *Foundations of Secure Computation*, R. D. Demillo *et al.*, eds., Academic Press, New York (1978), pp. 367–385.
62. MERKLE, R. C. Secure communication over insecure channels, *CACM* **21**(4), 294–299 (April 1978).
63. MERKLE, R. C., and HELLMAN, M. E. Hiding information and signatures in trapdoor knapsacks, *IEEE Trans. Inf. Theory* **IT-24**(5), 525–530 (September 1978).
64. MILLEN, J. K. Security kernel validation in practice, *CACM* **19**(5), 243–250 (May 1976).
65. MILLEN, J. K. Constraints and multilevel security, in *Foundations of Computation Security*, R. D. Demillo, *et al.*, eds., Academic Press, New York (1978), pp. 205–222.
66. MORRIS, R., SLOANE, N. J. A., and WYNER, A. D. Assessment of the National Bureau of Standards proposed Federal Data Encryption Standard, *Cryptologia* **1**(3), 281–291 (July 1977).
67. NEUMANN, P. G., *et al.* A provably secure operating system: The system, its applications, and proofs, Project 4332 Final Report, SRI, Menlo Park, California, February 1977.
68. PARK, M. C. Authorization with bounded propagation of privileges, M.Sc. thesis, Department of Computer Science, University of Houston, May 1981.
69. POPEK, G. J., and KLINE, C. S. Issues in kernel design, in *Proceedings of the AFIPS National Computer Conference*, Vol. 47, AFIPS Press, Montvale, New Jersey (1978), pp. 1079–1086.
70. POPEK, G. J., and KLINE, C. S. Design issues for secure computer networks, in *Operating Systems: An Advanced Course*, Flynn *et al.*, eds., Springer Lecture Notes in Computer Science No. 60, Springer, New York (1978), pp. 518–546.

71. Popek, G. J., and Kline, C. S. Encryption and secure computer networks, *ACM Computing Surveys* **11**(4), 331–356 (December 1979).
72. Popek, G. J., and Farber, D. A. A model for verification of data security in operating systems, *Commun. ACM* **21**(9), 737–749 (September 1978).
73. Rabin, M. O. Digitalized signatures and public-key functions as intractable as factorization, Technical Report MIT/LCS TR-212, MIT Lab. Comp. Sci., Cambridge, Massachusetts, January 1979.
74. Rabin, M. O. Digitalized signatures, in *Foundations of Secure Computation*, R. D. Demillo *et al.*, eds., Academic Press, New York (1978), pp. 155–168.
75. Reiss, S. P. Medians and database security, in *Foundations of Secure Computation*, R. D. Demillo *et al.*, eds., Academic Press, New York (1978), pp. 57–91.
76. Reiss, S. P. Security in databases: A combinatorial study, *Journal ACM* **26**(1), 45–57 (January 1979).
77. Reitman, R. P. Information flow in parallel programs: An axiomatic approach, Ph.D. thesis (also TR 78-350), Department of Computer Science, Cornell University, 1978.
78. Reitman, R. P., and Andrews, G. R. Certifying information flow properties of programs: An axiomatic approach, Technical Report TR 79-359, Department of Computer Science, Cornell University, 1979.
79. Rivest, R. L., Adleman, L., and Dertouzos, M. L. On data banks and privacy homomorphisms, in *Foundations of Secure Computations*, R. D. Demillo *et al.*, eds., Academic Press, New York (1978), pp. 169–179.
80. Rivest, R., Shamir, A., and Adleman, L. A method for obtaining digital signatures and public cryptosystems, *Commun. ACM* **21**(2), 120–126 (February 1978).
81. Roberts, R. W. Encryption algorithm for computer data manipulation, NBS, Fed. Register 40.52, March 17, 1975, 12134-12139.
82. Schlörer, J. Identification and retrieval of personal records from a statistical data bank, *Methods Inf. Med.* **14**(1), 7–13 (January 1975).
83. Schlörer, J. Confidentiality of statistical records: A threat monitoring scheme for on-line dialogue, *Methods Inf. Med.* **15**(1), 36–42 (January 1976).
84. Schlörer, J. Union tracker and open statistical databases, Report TB-IMSD 1/78, Institut für Medizinische Statistik und Dokumentation, Universität Giessen, June 1978.
85. Schroeder, M. D., Clark, D. D., and Saltzer, J. H. The MULTICS kernel design project, Proceedings of the Sixth Symposium on Operating System Principles, *Operating Systems Rev.* (*ACM*) **11**(5), 43–56 (November 1977).
86. Schwartz, M. D. Inference from statistical data bases, Ph.D. thesis, Department of Computer Science, Purdue University, W. Lafayette, Indiana, August 1977.
87. Schwartz, M. D., Denning, D. E., and Denning, P. J. Linear queries in statistical data bases, *Trans. Data Syst.* **4**(2), 156–167 (June 1979).
88. Shamir, A., and Zippel, R. E. On the security of the Merkle–Hellman Cryptographic scheme, Technical Report MIT/LCS, TM-119, MIT Lab. Comp. Sci., Cambridge, Massachusetts, December 1978.
89. Simmons, G. J. Symmetric and asymmetric encryption, *ACM Computing Surveys* **11**(4), 305–330 (December 1979).

90. Simmons, G. J., and Norris, M. J. Preliminary comments on the M.I.T. public-key cryptosystem, *Cryptologia* **1**(4), 406–414 (October 1977).
91. Snyder, L. Synthesis and analysis of protection systems, Proceedings of the Sixth Symposium on Operating Systems Principles, 1977.
92. Snyder, L. Theft and conspiracy in the take–grant protection model, Department of Computer Science, Technical Report No. 147, Yale University, November 1978.
93. Stonebraker, M., and Rubinstein, R. The INGRES protection system, in Proceedings of the ACM National Conference 1976, 1976, pp. 80–84.
94. Sugarman, R. On foiling computer crime, *IEEE Spectrum* **16**(7), 31–32 (July 1979).
95. Valiant, L. G. General context free recognition in less than cubic time, *JCSS* **10**, 305–315 (1975).
96. Walter, K. G., *et al.* Structured specification of a security kernel, in Proceedings of the International Conference on Reliable Software, SIGPLAN Notices 10.6, June 1975, pp. 285–293.
97. Weissman, C. Security controls in the ADEPT-50 time-sharing system, in *Proceedings of the 1969 AFIPS Fall Joint Computer Conference*, Vol. 35, AFIPS Press, Montvale, New Jersey (1969), pp. 119–133.
98. Williams, H. C. A modification of the RSA public-key encryption procedure, Rep. 92, University of Manitoba, Department of Computer Science, 1979.
99. Williams, H. C., and Schmid, B. Some remarks concerning the M.I.T. public-key cryptosystem, Report 91, University of Manitoba, Department of Computer Science, May 1979.
100. Winkler, P. M. Every connected graph is a query graph, Department of Mathematics, Emory University, March 1981.
101. Yao, A. C. A note on a conjecture of Kam and Ullman concerning statistical databases, *Inf. Process. Lett.* **9**(1), 48–50 (July 1979).

Index